KOREA AFTER THE WAR

My First Duty Assignment in Korea as a New 2nd Lieutenant

RUSSELL FLAGG

Publisher's Information

Author contact: flaggmgmt@msn.com

ISBN 978-1-953080-58-5

Cover design: Holley Flagg, Flagg Design

EBookBakery Books

Dedication

To my two platoon master sergeants in Korea who taught me everything about leading men. Thanks to Sgt. Temple and Sgt White, my 13 months in Korea was a rewarding experience.

Acknowledgments

To my wife, Holley Flagg, whose support and graphic design talents helped bring this book to life.

To my editor and publisher, I. Michael Grossman, whose gentle nurturing and publishing wisdom brought these personal Korean Army stories together into a comprehensive journal.

Table of Contents

My tour of duty in South Korea was at Camp Casey, 10 miles from the DMZ (demilitarized zone).

Preface

When I think of my Army experiences in Korea almost 70 years ago, the most memorable moments occurred when I least expected them. This book is about some of those moments.

As a new 2nd lieutenant with an ROTC commission, I had just completed my basic 3-month armored officer's course at Fort Knox. Korea was my first duty assignment. As a young officer, I had a lot to learn about the Army. How to take orders from officers at different levels above me. How to give orders to NCOs and GIs below me. How to manage those relationships in the field and at home base.

My 13 months in Korea provided a steep learning curve. I started by barely being able to manage a seven-man squad on a simple fence-building project. Nine months later I was commanding Company C with six tanks, six armored personnel carriers, 16 jeeps, a mess truck, a four-acre company area with 10 Quonsets, and 120 troops - half of whom were Korean who barely spoke English.

Life in Korea was enriched by the great friends of various ranks at the officer's mess. Those officers were my constant companions for 13 months - as we dined, drank, played cards or just lived together. Col. Stephenson and Col. Edmundson were two of those officers who influenced my tour of duty the most.

My 13-month tour of duty was with the 7th Cavalry Reconnaissance Battalion.

My Company B Commanders, Capt. Norbert, and Capt. Ramsey played an important role in my development, although they were both difficult to work for. Offsetting those officers were my two outstanding platoon sergeants, Master Sgt. Temple, and Master Sgt. White. Those two experienced soldiers were mentors who taught me by example how to lead troops.

The 7th Cav was based at Camp Casey, 10 miles from the demilitarized zone. While our mission was training, our once-a-month combat alerts were a sharp reminder that we were there in the event of a North Korean attack.

Our op area was in a valley five miles east of Camp Casey and the site of fierce battles five years before against the Chinese and North Korean troops. After those battles the Americans were able to push the North Koreans 10 miles north to the DMZ.

Our intensive training in the field during the week was offset by our relaxed off-duty time on weekends. That's when Col. Edmundson held his weekend poker games and Saturday night music fests, playing his Blue Grass guitar with a guest dentist from Seoul who played the five-string banjo at the Grand Ole Opry before his Army tour as a dentist in Seoul.

Col Edmundson transformed the 7th Cav officer's mess from a mundane dining facility to an enjoyable weekend retreat. I was fortunate to be one of the colonel's go-to junior officers, a regular at his poker games and one of his favored officers for special assignments. There were many ways to be successful in the Army, and I was learning one of them. I soon understood the value of being mentored by a senior officer like Colonel Edmundson.

It was a lesson I would take with me after my Army active duty was over. I enjoyed my experience in the Army. The same highs and lows during my tour of duty in Korea prepared me for a life journey many years afterwards.

This historic building in Seoul, was a visible reminder that Korea had a five thousand-year-old history.

GIs and Koreans shared the same valley as Camp Casey.

1

My Orders Read Korea.

After graduating from the Basic Officer's Leadership Course at Fort Knox in September 1958, my orders read Korea.

The Korean War was over and the armistice was signed in 1953. Korea was a peacetime training duty, although North Korea was always a threat. After 67 years, North Korea is still a threat, with ballistic missiles, nuclear weapons, and Kim Jung Un.

Korea was my first duty assignment, and I was lucky to have missed both the fighting in Korea five years earlier, and the fighting in Vietnam five years later.

In 1958, the Army had established bases throughout Korea with 25,000 troops stationed as a deterrence against North Korea. Camp Casey was my base and my new home for the next 13 months, It was 10 miles south of the DMZ. The camp was in a wide, fertile valley, with a main road leading 40 miles south to Seoul. The camp's strategic location in the valley was intended to block a main invasion route should North Korea cross the DMZ again.

We had monthly drills in preparation for a possible invasion, but our main daytime mission was to train. I was assigned to the 7th Cavalry Reconnaissance Battalion, one of the Army units stationed at Camp Casey. The 7th Cav was an armored unit with M-48 tanks, jeeps, and armored personnel carriers (APCs).

As a reconnaissance battalion, the 7th Cav's primary mission was to find the enemy. Our job was to make enemy contact with boots (and tanks) on the ground and to report our recon back to the main battle group.

Reconnaissance today is far different than it was 70 years ago. In the Ukraine war, both Ukrainian and Russian armies replace boots and tanks on the ground with drones, satellites, heat-seeking scanners, and remote systems to find and take out the enemy.

The 7th Cav had an illustrious history dating back to the Civil War. It's history included Indians wars in the West and General Custer at the battle of Little Big Horn.

Though there were no North Korean incursions while I was there, my 13 months in Korea had its highs and lows including cold weather, special assignments, and troops under my supervision that included recruits from the ROK (Republic of Korea) army.

Our 7th Cav officer's mess was a home away from home during our 13-month tour. The mainstay at the mess was an affable Korean bartender named Cheong, a 50-year-old ROK veteran who had seen real fighting against North Korean and Chinese soldiers. He also spoke Japanese from growing up in Korea during the Japanese occupation prior to 1945. Cheong was a gifted enabler of booze and good cheer, forever a bright spot after an arduous day in the field.

From the bare hills overlooking Camp Casey, our base looked like a giant array of half-cut beer cans.

The officer's mess, bar and kitchen were in an elaborate Quonset hut, as were all the other 7th Cav buildings including the battalion headquarters; our BOQs (bachelor officers' quarters); the four company's troop barracks, company headquarters, supply rooms, and maintenance buildings. Local Korean contractors made a successful business of transforming our half-round, corrugated metal buildings into comfortable living spaces and functioning Army troop facilities.

Camp Casey was a sprawling military installation, 40 miles north of Seoul and 10 miles south of the DMZ.

From the bare hills overlooking Camp Casey, our base looked like a giant array of half-cut-out beer cans. A wire fence defined our camp, separating it from the rice paddies in our valley.

A ramshackle Korean village, Geolsan-dong, sat 200 yards west of Camp Casey's main gate. There, 180 residents lived in mud-brick huts with thatched straw roofs. Most of the village men worked on the base, serving as cooks, laundry, and cleaning personnel, and provided tailoring and other essential services. The women in the village raised their families, washed clothes in the nearby stream and sent their kids to the village school.

A village named Geolsan-dong sat 200 yards west of Camp Casey. 180 residents lived in mud brick houses. The men worked for the Army at Camp Casey and grew rice and vegetables in the surrounding valley. Today this small village is a major Korean city with a population of 96,000, now called Dongducheon.

The men in the village worked at Camp Casey and grew rice and vegetables in the rice paddies and neighborhood gardens They were all part of the valley rice growing economy that existed side-by-side with Camp Casey.

Koreans and GIs shared the valley. In springtime, summer, and fall, the paddies and hillsides were cultivated by the villagers raising rice and vegetables. The valley and the neighboring valleys were also shared as training grounds and gunnery ranges by the Army.

In the 67 years since I was stationed there, the small village of Geolsan-dong has exploded into a thriving major Korean city. It now has a population of 96,000 and is called Dongducheon. Once mostly rice paddies, the city today has paved highways, stop lights, shopping malls, schools, a university, hospitals, homes, apartment complexes, and car dealerships.

As Microsoft Copilot reports: "The village and the base (Camp Casey) share a complex and intertwined history, reflecting the strategic importance of the region near the Korean Demilitarized Zone (DMZ)."

The heart of the 7th Cav was the motor pool. Each company had equipment that required daily maintenance for field operations.

2

Reporting for Duty

Reporting for duty at the 7th Cavalry headquarters on the first day of October 1958, I was met at the front desk by MSgt. Murphy, the sergeant major for the battalion. After a quick scan of my orders, he ushered me into the commanding officer's office.

Colonel Stephenson was at his desk and looked up with a smile. He was about 45, with a little grey hair, and was genuinely happy to see me. We exchanged salutes and I introduced myself.

"Nice to have you on board, lieutenant," he said.

After a brief summary of the 7th Cav, he assigned me to Company B, one of the three combat companies under his command.

"I'll call Capt. Norbert and let him know you are coming. He's your company commander and has been looking for a few officers for some time.

"You know, we're one of the nearest Army units to the DMZ in this part of Korea. This is peacetime duty, but there's always a tiger lurking just 10 clicks north of here. Keep that in mind when you're doing your tour, son," he said soberly.

"Now, on with your new assignment," he said brightly, "see the sergeant major outside. He'll tell you how to get to Company B. Good luck! Dismissed."

"Sir," and I saluted, and he returned my salute. I then spoke to MSgt. Murphy, who gave me directions to Company B, on the west side of Camp Casey.

After a 20-minute walk on a paved road to Company B, I found the company and the HQ's Quonset where I met the company clerk at the first desk. He directed me to the desk behind him where Capt.

Norbert sat, my new commanding officer. I saluted and introduced myself.

"Just got a call from the colonel about you, Flagg. Welcome aboard," the captain said.

Later that day at the officer's mess, I learned Capt. Norbert was OCS (officer's candidate school graduate) and a career officer. He had served in the Philippines as an enlisted man during WW II.

The captain was distracted with some paperwork and quickly came to the point. "We've been hoping to see some new lieutenants and we're happy you're here. I'm putting you in charge of the 2nd platoon that needs a lieutenant.

"As you can see, I'm a little busy so I suggest you go and find Sgt. Temple, who's been the acting 2nd platoon leader for the past three months. He'll fill you in. He's a master sergeant with combat experience in Korea, so pay attention to what he says. He'll get you started. Come back to the HQ here at twelve, and we'll go to lunch at the officer's mess in my jeep. That's all. Dismissed."

After "sir," I saluted and left. I was a little disappointed that the captain didn't have more time to tell me what my new duties were. After leaving the HQ, I saw a sergeant coming out of the 2nd platoon barracks across the street. He was 6 feet tall, had a well-trimmed mustache, wore starched and freshly pressed fatigues with master sergeant's stripes, and looked like Clark Gable. That was Sgt. Temple.

"Hi, sergeant," I said, meeting him on the company street, "I'm Lt. Flagg, and the captain told me to see you about the 2nd platoon. He's assigned me as your new platoon leader, and said you'd fill me in."

Sgt. Temple gave me a quick look and assumed this was my first tour. But without hesitation he offered me a gracious welcome, "Glad you're here, lieutenant. Captain said there'd be some officers like you comin' along. Just to bring you up to date, I've been inspectin' the barracks and we're havin' trouble with Katusas and their boots," he said matter-of-factly.

"What are Katusas?" I asked.

Almost 50% of the troops in the 7th Cav were Katusas (Korean Augmentation to the U.S. Army). and were young Korean Army recruits, conscripted from rural villages with no military or mechanized equipment training. Most did not speak English.

Sgt. Temple laughed. "You'll learn soon enough.

"The short version is they're 'Korean Augmentation to the U.S. Army,' that's K.A.T.U.S.A. The long version is that they're young Korean lads comin' off some rice paddy with no military experience. They're paid by the ROK to serve in our Army. We outfit 'em with uniforms, boots and all the gear we provide to our own GIs.

"We have 30 troopers in our platoon and almost half are Katusas. Our unit is mechanized and none of these guys have ever driven a car," he said shaking his head for emphasis.

Still confused, I asked, "Do they speak English?"

He laughed again, "You'll find out."

It was clear after a few minutes he knew more about everything in Korea and the Army than I did. He was a career soldier, had fought in Korea, and I immediately realized how much I needed him as my platoon sergeant. After our chat, I was so taken by Temple, I instinctively started walking with him when he turned to leave.

"Where are you going, sergeant?" I asked.

"Goin' to lunch at the mess hall. Don't you lieutenants eat lunch?" he said with a smile.

"Oh, of course," I said, somewhat embarrassed, "Yes, I'll be going to lunch with the captain. But I hope we can catch up after lunch."

"No prob," the sergeant said. "I'll be in the motor pool. It'll be a good time to show you the vehicles and meet the men. See ya' then."

I walked back to the HQ and met the captain and 2nd Lt. Fitzgerald waiting for me at the captain's jeep to go to lunch at the officer's mess. Fitzgerald was an old hand and had been in the company six months as the 1st platoon's leader. Kim was the captain's jeep driver, and he drove the three of us to the officer's mess, about five minutes away on the same paved road I had walked earlier. The captain and all the officers regularly rode back and forth to the officer's mess, which provided all the officer's meals when they were not in the field. The officer's mess was on a hillside about a half mile from Company B and 200 yards up the hill from the 7th Cav HQ. The officer's BOQs (bachelor officers' quarters) were also located on the same hillside. All the buildings were Quonsets.

The officer's mess was not only the dining facility for all the officers but also the social and drinking retreat during non-duty hours. When we were not in the field, we were at the officer's mess. During lunch I kept thinking about Sgt. Temple, and I asked Lt. Fitzgerald what he knew about Temple.

"Sgt. Temple had a battlefield commission and was promoted to major during the Korean War. The Army really needed combat officers back then with the officer corps getting killed off left and right. He commanded a tank unit in some fierce fighting up north. He was a 'mustang,' an enlisted man with a battlefield commission, and must've have been a helluva combat soldier.

"After the War, the Army cut budgets and reduced their entire officer ranks. Sgt. Temple was a casualty because he never finished high school.

"As a mustang, the Army gave him two choices. He could leave the Army with no opportunity for a 20-year career. Or he could stay in the Army with the rank of master sergeant and have a career and retirement benefits after 20 years. Temple chose the Army," Fitzgerald explained.

I was fortunate indeed to have Sgt. Temple!

After lunch, I met Sgt. Temple and the 2nd platoon working in the motor pool. The motor pool is the heart of any mechanized unit. Our platoon had two tanks, two APCs, and five jeeps, and a formidable fighting force, but a lot of vehicles to maintain on a daily basis.

After the Korean War, the Army's newest equipment was shipped to Europe to combat Russian troops and the Iron Curtain. The Army in Korea was left with some of the same equipment that fought in the war five years before. Old equipment like ours needed constant repairs. Our time in the motor pool was spent keeping old vehicles running.

Mechanized units in Korea had one additional problem. With an agreement between the U.S. and the ROK, the ROK provided supplemental troops to the U.S. Army to defend against North Korea. Those troops were Katusa recruits with no experience operating jeeps, tanks, and APCs. So Katusas assigned to us were relegated as mounted infantry and riding in APCs. Very few Katusas ever learned to drive a jeep, let alone a tank.

When I met Sgt. Temple in the motor pool, he explained more about the missing boots issued to Katusas.

"Lieutenant, the GIs and the Katusas are all issued two pairs of boots, two sets of fatigues, underwear, coats, helmets, webbing, sleeping gear - everything they're needin' for duty. "Our Katusas are comin' from rural villages and never have two sets of anything. An extra pair of boots to them was found money, and they can't wait to go to the village and sell 'em. The ROK pays them almost nothing and that extra won (Korean money) for the boots really has value.

"The 8th Army has an annual inspection and both GIs and Katusas must show every single piece of Army-issued gear. That includes two pairs of boots.

"If the colonel hears about any missin' boots, there's hell to pay, lieutenant. We daresn't let that happen. So I've ordered the Katusas who sold their boots to go back to the village and buy 'em back. If they don't have the money, we help 'em out, but we gotta have two pairs for every trooper," he said.

I realized being a platoon leader was a full-time job just providing basic training to the Katusas. Sgt. Temple thrived in that job, and I realized what a steep learning curve I had just to catch up.

I was also beginning to understand the rhythm of a 12-hour workday. Each day began at 5:15 with a wake up call. At 5:45 a jeep arrived to take one of us lieutenants to the company to stand reveille at six. Reveille started the day with the whole company standing in formation in the company street to take roll call, Monday through Saturday, rain or shine, summer or winter.

Reveille was at 6 am and the start of the 12-hour duty day, when the whole company stood in formation on the company street each morning to take roll call.

After reveille, the troops had breakfast and returned to the barracks to make beds and put their barracks in order. At eight, its motor stables (an old horse-cavalry term) in the motor pool, followed by classroom work or training in the field.

In addition to my responsibilities as the 2nd platoon leader, the captain called me to his office the first week and said he was assigning me as his new company supply officer.

"Lieutenant, go find Sgt. Tripp in the supply room. He'll explain your new duties." I was beginning to understand how the captain delegated.

Sgt. Tripp was the company supply sergeant, in charge of all the company's supplies, from gas masks to M-1 rifles. He was required to keep accurate records of every item in his care. Tripp spent most of his time in his Quonset, as was evident from the body odor I noticed when I walked into his hooch.

Each Quonset barracks had two rows of bunks facing a center aisle with a clothing rack against the wall and a foot locker next to each bunk.

Tripp was about 33, a staff sergeant in the Army for 10 years. I found him at his desk, a little hung over, and not fully shaven. I introduced myself. Without standing or bothering to salute, he established his supply room prerogatives right away. "Lu-tenant, I gotta lot 'a work to do, and don't have much time for chit-chat. What's on your mind?"

"The captain assigned me as your new supply officer. He said you'd explain my new duties to me," I said.

Sgt. Tripp rolled his eyes. "Lu-tenant, the captain's full of surprises. Ya 'know sir, I've been in the Army 10 years and still don't know everything there's to know about supply. It'll probably take me another 10 years to explain it to you," he said with a guffaw.

"Truth is, it's complicated business. See these records here?" he said, reaching for some supply records from an open file drawer in his desk. "Every one of these records must be kept accurate and up to date. I spend hours on 'em just so they'll pass inspection. "I just don't have time to start teachin' you 'bout supply," he said resolutely, "if I'm gonna be losin' my stripes doin' it."

Teaching me how to be a supply officer was not his problem. After no further conversation, I left Tripp's supply room. Walking back to the motor pool, I thought of all the things I could have said to Tripp. I didn't want to start ordering a senior NCO around, and I also didn't want to get in trouble with the captain. It was a good leadership problem – *how to deal with a noncom under my command who was not responsive to me without getting in trouble with my commanding officer.*

After six visits to Tripp's supply room - three a week for two weeks, Tripp and I reached a détente. I wasn't expecting Tripp to give me any help being his supply officer. Yet visiting Tripp in his supply room was like visiting a sick patient in the hospital: it wasn't going to change anything, but the patient had to acknowledge I was making the visit.

By doing nothing, I was beginning to understand how the army worked. Tripp would continue to do whatever he was doing undeterred by me as his supply officer. I would continue as his de facto supply officer. The captain would continue with his busy schedule and never notice either one of us.

When our tanks, APCs (Armored Personnel Carriers), and jeeps took to the field, they made full use of every streambed, mountain pass, and rocky or open terrain.

3

Field Op

Captain Norbert called a meeting of his platoon leaders on Friday morning of the third week I was in Company B.

"Gentlemen," he said, "we're going on a Field Op (Field Training Exercise) Monday, and I want you all in the motor pool today to make sure your vehicles are ready for Monday. Questions? Dismissed."

Sgt. Temple was already in the motor pool when I met him. He knew about Monday's Op and had the crews working.

"Lieutenant, I've been checkin,' and we got some vehicles that ain't goin' nowhere Monday," he said.

"One jeep's out with a wheel bearing. Another needs a clutch. One of the tanks is spittin' black smoke and the battalion ordered it back in the shop for a valve job. Everything that's runnin' needs maintenance. We got all the GIs workin' who aren't on sick call.

"The Katusas are in the motor pool too. They ain't much good, but they can sweep, wipe down jeeps and knock mud off the tanks and APCs. We gotta start teachin' them something."

2nd platoon had 17 GIs and 13 Katusas. Most of the Koreans were nice young men, but the 7th Cav was their first Army experience, and military discipline and American culture was all new. By contrast, our GIs had spent time in the Army, were driving jeeps and tanks, and knew their way around. The differences were even more obvious with GIs and Koreans sharing the same barracks and bunking side by side with no separation.

The Army's attempt to integrate Americans and Korean troops may have been well intentioned but was a hilarious experiment at times. Some fun-loving American GIs, for instance, took it upon themselves to teach their Korean bunkmates English. The first words were always swear words. I overheard a Katusa stub his finger and exclaim, "Somma fuckin' bitch."

My Korean jeep driver was the exception. Chou was college-educated, spoke English, and came from Yongsan, a suburb of Seoul. He had been driving jeeps for a year and never had a serious accident. But mountain passes and valley streambeds were his nemesis.

The DMZ was an active border because North Korea constantly engaged in feints and other at-border alerts.

In Korea, every mountain pass or valley streambed was a military access road in wartime. When I was there, almost all Korean roads were just earthen tracks, muddy in winter, dusty in summer. We never traveled a single paved road anywhere in Korea except at Army bases and in downtown Seoul.

The platoon worked all day Friday and by four, the six vehicles still operating were good-to-go. During that time, Sgt. Temple also checked the vehicle radios. He chuckled. "Some guys turn off their radios goin' to the Op. They've been here for months and know the trainin's always in the next valley five miles west of here. Those guys don't need radios to get there.

"But before we go Monday, I'm askin' those guys nicely to keep their radios on as a personal favor to me. I don't want the captain raisin' hell with us 'bout radios bein' off."

I was learning how to be in command without issuing orders. We finished at four Friday, and the troops went back to the barracks to wash up and have dinner. The officers rode to the mess for drinks and dinner.

The 7th Cav officer's mess was a lively place after four, like a country club bar after a match. Rank had its privilege, and captains did most of the talking, with the colonel chiming in on occasion. Lieutenants, unless they had been there a while, were not part of the inner circle.

The officer's mess was unique. It was a Quonset that was a dining facility but also a social club. All 35 officers in the battalion ate, drank, played poker, shot pool, or socialized there, like a men's business club.

The officers at the mess represented every level of Army experience. Some were West Pointers, some OCS (officer's candidate school), most of the young officers were ROTC. Some older officers served in World War II or the Korean War. Others served in Europe, the States, or the Pentagon. One was an Army aviator, and another was a nuclear physicist. There were 16 captains with 10 years or more years in the service. One major had 13 years, and our commanding officer, Col. Stephenson, was a lieutenant colonel with 16 years of service.

In that hierarchy there were 17 relatively new 2nd lieutenants, mostly ROTC. My fellow lieutenants in Company B were Brian Fitzgerald and Tom Clarkson. Both were ROTC and Korea was their first duty assignment. Brian had been there six months and was in the inner circle. Tom was a newbie and arrived a week after me. But Tom Clarkson was special.

Tom was a professional baseball player. He had been drafted by a major league team from college and had been playing minor league ball before his two-year Army hitch. The regimental commander at Camp Casey, a lieutenant general, knew a professional baseball player was coming to his regiment. Even before Tom arrived, the general had him on the regimental baseball team as a right fielder.

Captain Smithfield was an HQ staffer and an avid baseball fan. He often cornered Tom in the bar to talk baseball. It was clear Tom didn't mind fans like Smithfield. Korea was just another detour in his baseball life - one he called his 'Korean farm team.'

Though Tom was a baseball celebrity, at the 7th Cav he was just another junior officer when it came to sleeping quarters. All 2nd lieutenants were bunked in the same unadorned BOQ (basic officer's quarters) in a Quonset 100 yards up the hill from the officer's mess. Our Quonset was exactly the same as the enlisted men's barracks with an open interior, a double row of bunks, a footlocker and clothing rack with no partitions between bunks. There was no indoor toilet. A wooden urinal was located immediately outside the BOQ's front door. The only flush toilet was down the hill in the officer's mess, behind the bar.

Korean workers from the village provided daily housekeeping, laundry, and dry cleaning. They also took care of the captain's and major's BOQs which were more elaborate - with room partitions and an indoor toilet. The colonel had his own hooch with all of those amenities.

Camp Casey had a PX, a movie theatre, an NCO club, and a craft shop for photography and other hobby crafts. There were no women on the base. American Red Cross ladies came to the base twice a month to entertain enlisted troops with good cheer, and donuts and coffee.

The 7th Cav officers mess was not a stop for the Red Cross' donut dollies. Our good cheer was more of a self-generated kind, knowing we would all be living together for our 13 month's tour. The duty schedule for Saturdays was half a day for both troops and officers. At the officer's mess, it was usually a one o'clock lunch followed by a relaxed afternoon or a regular poker game. While the poker stakes were low and a player might lose $10, the competition was fierce. As a new lieutenant, I always welcomed a senior officer's complaint. "Flagg, I never expected three aces from you. You may not know a thing about the Army, but you're an SOB at poker."

Poker was one game I was good at, offsetting my other deficiencies at the 7th Cav.

Sunday was a full day off for troops and officers. We could sleep in, catch up on letter writing, read, or follow other recreational activities. Sunday (and Saturday afternoon also) was off duty for everyone except the OD (Officer of the Day).

The OD was the one designated officer in the 7th Cav that was on duty every weeknight and each weekend day and night when the HQ and company HQs were off duty. The OD's tour of duty usually started at four pm, continued until six am the following morning and included the control and security of the entire 500-man battalion. 2nd lieutenants were the only ones assigned as ODs and then on a rotating basis once or twice a month.

The OD had use of a jeep and driver to patrol the entire 7th Cav area and inspect troop barracks, motor pools, and HQs. ODs were required to be in uniform, armed with a loaded .45, and to be alert and on call during their tour. The OD was responsible for all four companies inside a mile area of the Camp Casey compound, and during my first OD tour, I had several memorable incidents: a medical emergency, a drunken brawl, and I caught a Korean prostitute slipping into Company C barracks of to sell her services. We drove her to the MP (military police) post where she spent the night and was released the next morning to go back to her village.

The OD's duty was always daunting because they were challenged to keep order and were required to make unaided decisions that affected the whole battalion. While we had guidelines, there were always exceptions.

In one instance, during another OD's experience, a young soldier, Pvt. Canales, got drunk and went AWOL (absent without leave). He'd been told that his Korean prostitute girlfriend was cheating on him, and he went to a nearby village with his loaded rifle. The

OD was alerted that Canales was AWOL and left with his weapon. He immediately radioed the MPs who drove him to the village to apprehend Canales. They found him drunk, threatening to kill his girlfriend and shoot up the village. Fortunately, he was arrested before he did any damage. Canales was court marshaled and sentenced to six months in the 8th Army stockade in Seoul.

No one was hurt, but it was the kind of scare that occurs when 500 soldiers are living together and have the potential to do damage. Soldiers are expected to manage their emotions and keep their anger in check, but there are always exceptions. That night, the OD had the harrowing task of preventing a shooting and a village rampage.

My first weeks in Korea were a blur of those kinds of Army experiences. I was concerned about what I didn't know about the Army. At the same time, I was also adjusting to life as a young officer in Korea.

MONDAY MORNING: MOVE-OUT FOR THE FIELD OP

"Radio check. All platoon leaders must be on channel one. Let's hear from you," the captain instructed. One by one each of us responded.

The company's command channel was one. My platoon channel was three. The other platoons had their own channels.

"Move out," the captain barked on one, and I repeated the command on three. Two jeeps, a tank, and two APCs belched exhaust and were on the road following my jeep.

"Keep your intervals. HQ wants 10 yards between vehicles," the captain radioed. We were all aware that Capt. Townsend, the S3 (operations officer) was following us in his jeep. I repeated the order.

Lt. Fitzgerald had warned me about Capt. Townsend the night before. "Watch out for Capt. Townsend. He'll be following us Monday and he's a bear for proper intervals between vehicles."

"Townsend is on a crusade for 7th Cav combat readiness. He has no problem countermanding captains and stopping vehicles if they don't follow his 10-yard interval," Fitzgerald said. "He's the same rank as the captains but he's top dog in the field. He represents HQ and the colonel."

On Monday morning, after following us for a mile, Townsend stopped our column. Our vehicles were not 10 yards apart.

"Why the hell are we stopping, lieutenant?" Sgt. Galant yelled, riding in the back of my jeep with his crewmate, Cpl. Baker. Their jeep was in the shop, and they were riding tandem with us.

"Capt. Townsend is S3 and calls the shots. He works for the colonel, and we follow his orders," I yelled back.

"Hey Sgt Galant, you and Baker better hold on. I don't want you guys falling out," I yelled, ending the conversation. I didn't want to start a gripe session about Capt. Townsend.

"Y'sir," the sergeant said grumpily.

We were following Fitzgerald's 1st platoon, and his vehicles were completely hiding the road with their dust. "Keep 10 yards," I yelled to Chou over the road noise. He was leaning over his steering wheel, trying to see the 1st platoon's tank in front of our jeep. The windshield was down but he was still having trouble seeing the tank. "Chou, slow down," I yelled. He mumbled "Y'sir," still leaning forward.

The valley to the Op had been a battlefield five years before. Off the road, we could still see traces of the war in the valley and in the rolling hills, 1,500 feet above. Those hillsides were surprisingly green, unlike the valley. Their base was strung with concertina wire with posted signs that read: 'MINES.' in Korean and English.

There was no cultivation in those hills where the uncharted mines were. Those pristine hills had become lush green preserves for trees and plants supporting all kinds of animal and bird life. We often saw wild boar, roe deer, and game birds in the green hills from the road. Some GIs even hunted pheasants there, carefully treading on well-worn pathways to avoid mines. Village papa-sans regularly climbed the hills, collecting firewood for their village huts.

5 years after the war there were still thousands of mine fields throughout the country, identified by signs and concertina wire. They became green preserves supporting robust trees, plants, and a wide array of bird and animal life.

When we arrived at our Op site, it was readily apparent that the pockmarked terrain had also been a battlefield. There were bomb craters, trenches, and tank defilades (dug-out pits to lower a tank's profile) everywhere. It was an excellent training area because it was a real battlefield where American, North Korean, and Chinese armies had fought years before.

The valley was 10 miles long running north to south. It was a major attack avenue for the North Korean and Chinese armies advancing from North in the beginning of the war. It was similarly an attack avenue when the American armies advanced from the south, pushing the enemy all the way up to the present DMZ. There were no local inhabitants or cultivation in the valley because the Army had exclusive use for training and live fire exercises.

After the three platoons arrived at the base camp, the captain assigned the 1st platoon to the north, simulating a North Korean attack. He then assigned the 2nd and 3rd platoons to the south, defending against the attack. Following the captain's radio orders, the three platoons were set in motion. Each platoon's radio channel was alive with: *"Move forward," "Move back," "Hold in place," "Can't you read the map?" "Don't stop on the hill, you're exposed to enemy fire."*

The valley was a 10-mile chess game with vehicles moving in all directions on offence and defense. But the humans driving the vehicles were not chess pieces. Mistakes were made, orders were ignored, and corrections were required. It was a training exercise.

My platoon was constantly in motion because Capt. Townsend had picked my platoon to monitor. He was with me all morning, correcting me and ordering my jeeps, APCs, and a tank all around, and re-adjusting every location of our 2nd platoon deployment. At one point, he drove by and told me in person, "Flagg, I've never seen a more poorly commanded unit than yours. Your vehicles are on the hills and constantly exposed to enemy fire. Lieutenant, you're not doing your job."

After a morning with Townsend, it was a relief when the captain radioed, "Platoons one, two and three: return to home base. Lunch and a briefing." Our company's mess sergeant, Sgt Dempsey, had his mess truck and crew at base camp set-up and ready when we rolled in. One by one, each platoon was in the chow line for creamed chipped beef on toast, coffee, and short cake. The troops called chipped beef on toast 'shit on a shingle,' or SOS, but in was surprisingly good after a frustrating morning.

"Gentlemen," the captain began his briefing as we ate. "We'll have to do better this afternoon. Capt. Townsend's report to the colonel is not good. Flagg, you got to keep your vehicles off hills. Fitzgerald don't move your scout jeeps so far in front. Clarkson, keep your command channel on. I had trouble reaching you.

"All your crews must avoid exposing the sides and backs of your vehicles to the enemy fire. If it were a live war, they'd all be dead."

The afternoon went about the same as the morning. My jeeps, tank, and APCs were still not chess pieces, and their drivers still made mistakes, didn't follow orders, or were challenged by difficult terrain. By 3 pm, the captain halted the north/south battles and ordered the units back to base camp for a final briefing.

"Lieutenants, this Op was not up to our company's standards. You've got to do better. Let's at least make sure the trip home follows Capt. Townsend's 10-yard intervals. I don't want Townsend ripping us again for improper intervals between vehicles. Keep your radios on. Let's make a good showing. That's all. Dismissed."

The trip back was dusty, slow, and uneventful. Capt. Townsend followed our column back to Camp Casey but didn't stop us once. We assumed he already had enough material in his report for the colonel without needing more.

We arrived at B Company's motor pool about 5:30, and spent another hour refueling vehicles, making notes of maintenance to be done, and returning weapons to Sgt. Tripp's supply room. At 6:30 the troops were dismissed to wash up for dinner, and the officers drove to the mess for a late dinner.

It was a long day. I found Townsend's interruptions and scolding irritating and frustrating. I expected my first Op to be hard, but I didn't expect so much human error or vehicle limitation.

It was going to be a long 13 months.

During my tour, almost every road in Korea was just an earthen track, muddy or frozen in winter, dusty and rutted in summer.

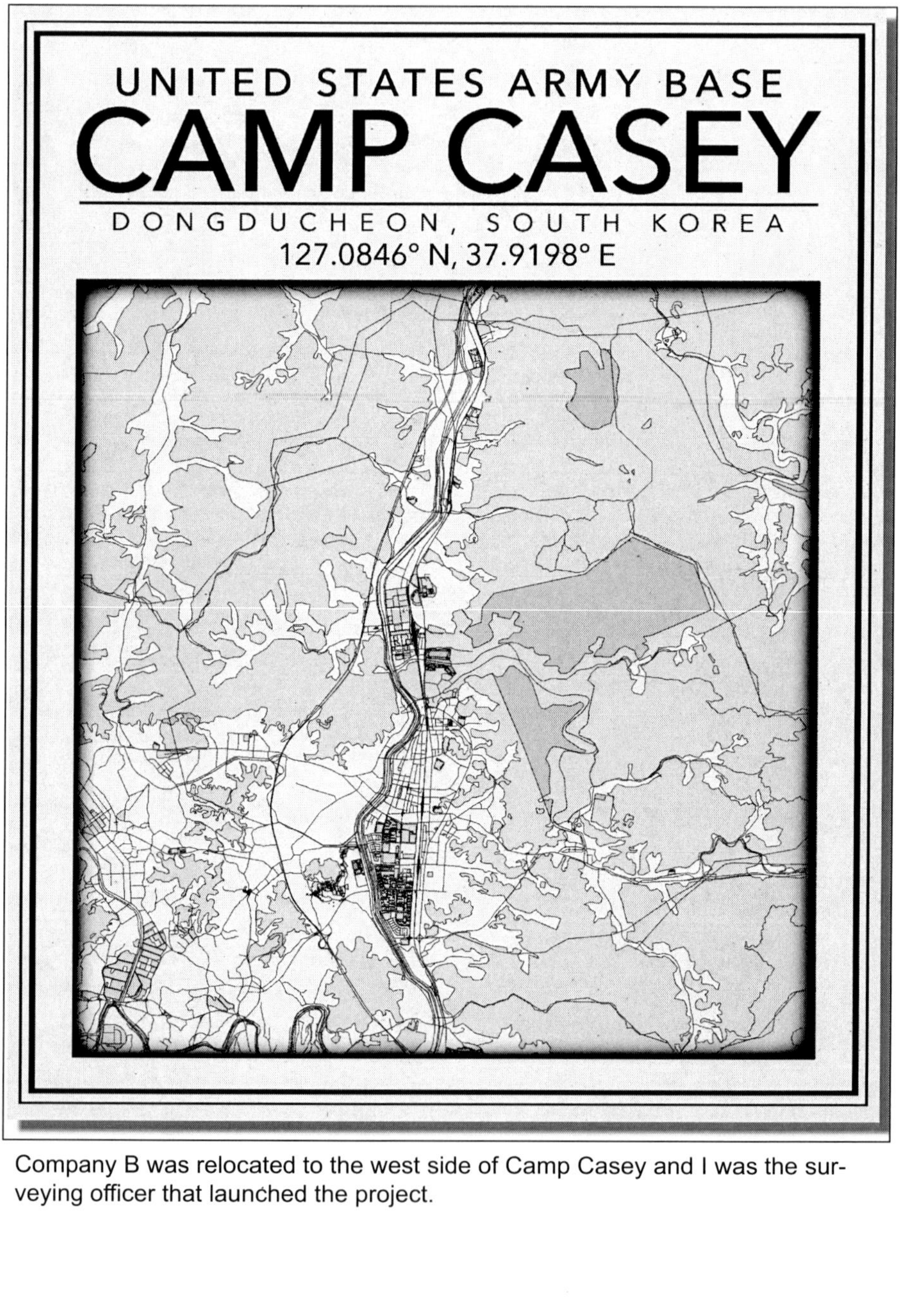

Company B was relocated to the west side of Camp Casey and I was the surveying officer that launched the project.

4

Surveying a new site for Company B

"Lieutenant, we need you to survey a new site for Company B," Col. Stephenson announced.

The colonel had asked Capt. Norbert to send a lieutenant to HQ for a new assignment. The captain sent me, having arrived at the 7th Cav just weeks before, on October 1.

Col. Stephenson began by showing me a map of Camp Casey on his desk. He pointed to where Company B was and where it had to be relocated. He said my job was to lay out the perimeter in the new area and erect a barbed wire fence along that line.

Company B had been on the east side of Camp Casey for five years, ever since the Korean War. Most recently, the east side was being taken over by a new artillery unit with Honest John rockets. Company B had to move.

"You seem competent enough, at least for this job, Flagg. I want you to start tomorrow morning and report back to me by the end of this week. I need to see what you're doing. I'll let Capt. Norbert know this assignment has my highest priority. We got to move the Company B before the snow flies. That's all. Dismissed."

"Sir," and I saluted, and left his office and got a ride to Company B to report to the captain about my new assignment.

"Now Flagg, do not get carried away with this assignment. You still got work to do here," Capt. Norbert began. "Talk to Sgt. Temple, and let him know what you're up to. Take one of your squads tomorrow, and do what the colonel wants, but don't forget you work for me. Dismissed."

"Yes sir," I saluted. The captain left no doubt who I worked for.

I met Sgt. Temple on the company street and told him about my new job.

After reporting for duty at Company B, the captain sent me to Col. Stephenson at the 7th Cav HQ for an assignment to survey a new site for Company B and to erect a barbed wire fence. This led to my training in managing a 7-man work detail.

"I kinda figured that rocket company was goin' to make waves," Sgt. Temple said. "With their 24-hour armed guards, floodlights, concertina wire, and not a peep about the rockets, I knew they were the real thing. They got A bombs. Glad we're movin'."

Temple always knew what was up.

"The captain told me to take one of our squads tomorrow to begin laying out the new company area. Which squad should I take, sergeant?" I asked.

"Take Sgt. Nunez's, sir. We need the other squads to work on vehicles. But just remember, Nunez is Nunez. You'll have manage him and his squad. You can do that, can't you, lieutenant?" Temple said with a little chuckle.

We both knew Sgt. Nunez was excitable and often confused when following orders. But he was a goodhearted guy and was well-liked by his squad and the other NCOs. Nunez had enlisted about eight years before and was a career soldier.

Sgt. Temple added, "You know, lieutenant, Nunez is always goin' to be a work in progress. So make sure you explain your orders to him carefully."

Sgt. Temple didn't say it, but after watching me for three weeks, he knew I was also a work in progress. (I suspected Sgt. Temple thought a seven-man squad was a good place for me to start learning about leading a work detail.)

The next morning, I was the reveille officer for the company's 6 am reveille. Afterwards, I told Sgt. Nunez to report to me on the company street at 8 am with his seven-man squad. I had my jeep and driver Chou ready and one other jeep lined up to carry his squad and our equipment to the new site.

At 8 am, I waited on the company street - but no Nunez. After five minutes, Nunez came running, distraught and accompanied by Pvts. Jones and Smith, two members of his squad.

"Where is the rest of your squad, sergeant? I asked, trying not to act upset.

"Lieutenant, sir, I had the whole squad this morning," Nunez said, still out of breath, and with his eyes looking down to avoid my searching look. "But I didn't know Jacobi and Allenby would be going on sick call this morning. I also forgot Carter and Watson were coming off guard duty last night and had the morning to sleep. The first sergeant has Kim in the motor pool painting."

"Sgt. Nunez, the colonel gave me a job, and I need your seven-man squad," I said, probably explaining more than I needed to.

"Sir, my squad is always ready for duty" Nunez said apologetically, "jus' not today."

I was remembering Sgt. Temple's comment, "Nunez is Nunez." Except for Nunez, I had everything set to go: a 50-meter tape measure; post hole diggers and shovels; HQ company's 12-foot fence posts, and jeeps at the ready.

I also arranged an earthmoving tractor and driver from HQ to clear the trees and shrubs from the line that would define the company's new perimeter. I'd had everything I needed except Nunez's seven-man squad. I was beginning to understand how the Army worked at the squad level.

With Nunez's skeleton crew, and with the tractor and driver, we began the first day on the west side of Camp Casey. The time was well spent defining the entire area of the new Company B site. The HQ tractor and driver were busy all day clearing a path for the fence line that I had laid out. Nunez and his two men spent the day clearing brush from the path the tractor cleared.

The next morning, I was reveille officer again. Afterwards, I again told Sgt. Nunez to have his seven-man squad ready on the company street at 8 am. This time Nunez did better, arriving with Pfc. Carter, and Pvt. Watson, Pvt. Jones, and Pvt. Smith. Kim was still painting, and both Jacobi and Allenby had contacted VD (venereal disease) in Seoul and the medics were keeping them an extra day for observation.

On the second day with Nunez and his four-man crew, we made progress and cleared 400 yards of brush and small trees with the help of the HQ tractor and driver. They also planted stakes where the fence posts would be dug, and the barbed wire would be strung.

"Have you done fence posts and barbed wire before sergeant?" I asked Nunez as we were about to set the fence posts.

"Oh certainly, sir," Nunez replied.

"Well, that's good sergeant because I have to go to the captain's meeting at 10, and I'm going to leave you in charge. Dig the post holes 10 feet apart, four feet deep, and use the tape measure. Make sure the fence posts are upright and uniformly eight feet above ground.

"Anchor the posts with rocks to keep them upright. I be back in an hour or so. Make sure the posts are 10 feet apart, four feet deep and vertical. You got the order, sergeant?"

"I got the order, sir," Sgt Nunez said brightly.

His squad started placing the fence posts as I drove off with Chou for the captain's meeting.

At the meeting, the captain said the colonel phoned him and said the company's move had the highest priority. The new camp must be completed before the snow flies. It was the same message I had given the captain three days earlier.

I was beginning to understand the chain of command. If the captain received an order directly from the colonel, it had greater importance than if a lieutenant relayed the order. Lieutenants were just a clog in the smooth transmission of orders.

After surveying and clearing the fence line of the new company area, the 7-man squad erected 800 yards of fence and strung barbed wire.

After the captain's meeting, I drove back to my fence project. On arriving, I saw that Nunez's squad had set three posts in the ground and had begun to set the fourth. What was obvious was how ineptly Nunez had set the posts. They were set helter-skelter at different heights, pointing in different directions.

"Sergeant, I'm sorry, but it looks like the posts are not uniform and will have to be reset. I thought you said you had done this work before?" I asked Nunez as we stood away from his squad. I didn't want to embarrass him in front of his men.

"Sir, I did fence posts years ago before I got my stripes. I told Carter the posts should be more vertical. We'll need a level to set the posts upright sir," Nunez explained defensively.

"Okay, okay sergeant, we're all learning. Take a break, then re-do the fence posts so they are exactly vertical and the same height. We'll bring a level tomorrow, but today you'll have to do it by eye, okay?"

"Yes sir," Nunez said dejectedly, and walked back to the squad. I overheard him grumbling, "Carter, I told you to make them vertical."

I took a deep breath and tried to think what Sgt. Temple would do. I looked at the 600 yards of the fence line still to be cleared and realized I had to direct Nunez and the seven-man squad, while making it appear that Nunez was leading the project. I was also hoping that Pfc. Carter, a bright young man, would step up to help Nunez, but it was still my job to take charge. My officer's leadership course at Fort Knox didn't cover leading a seven-man squad to set fence posts.

By the first week in November, with the captain and the colonel overlooking the job, the fence line was finally looking uniform with the fence posts 10 yards apart and exactly vertical, and by the 10th of November, we began to string barbed wire.

The whole area within the defined fence line perimeter was also being transformed. The new Company B area had been laid out and concrete poured for the foundations of the 10 Quonset huts that would serve as Company B's barracks and as ancillary buildings for 120 troops.

The Quonsets would become the troop's barracks, a mess hall, a supply room and a headquarters office. A large shed was constructed to serve as the covered motor pool area. Water and sewer lines were laid, septic tanks were buried, electric and telephone lines were strung and a new asphalt road was laid for the company street.

Company B was humming. Troops were working in the Quonsets and on the grounds every day during the rest of November and the beginning of December to complete their new company home.

By December 10, Company B officially moved into its new area. It had all started with my fence line project in October. I was learning how to lead a seven-man squad.

The relocated Company B site had 10 newly constructed Quonsets to accommodate the 120-man company. These half-round buildings were quickly constructed by highly skilled Korean contractors.

The living and working quarters of every company were Quonset huts, with barracks, a mess hall, a supply room, and a company headquarters. They had comfortable, insulated interiors with heating and lighting systems, excellent flooring, doors and windows.

Christmas in the Army in Korea was just another day. While the Army provided frozen turkeys for the mess halls, Christmas that Thursday was just a day off unless you had guard duty.

5

Christmas in Korea

Christmas in the Army in Korea was just another day. There were few Christmas trees, fewer holiday lights, and no furloughs between Christmas and New Year's for the troops. While the Army provided frozen turkeys for the mess halls, Christmas that Thursday was just another day off unless you were on guard duty.

Col. Stephenson, called me to his office two weeks before Christmas. He had learned of my artistic ability while I was working with him on the fence job for Company B's new area in Camp Casey. "Flagg, you did okay with the Company B fence job I gave you. I have another one for you today. We need to dress up our battalion headquarters area for Christmas.

"Do you see that old M-46 ceremonial tank sitting in front of the HQ? We need to dress it up. I wanna see some fancy reindeer in front, pulling it like Santa's sleigh. You said you're good at art the last time we talked. I'm putting you in charge of making a pair of reindeer for the front of that tank. I want it by Sunday. We need some Christmas cheer around here. Can you do it?"

"Yes sir. (Pause) But I'll need some time Friday at Ralph Nordeck's craft shop to make the reindeer. Is that going to be okay with Capt. Norbert, sir?"

The colonel rubbed his hand over his grey hair. "You're asking a lot for a couple of reindeer, Flagg," the colonel said with a little smile.

"Okay I'll call Norbert and tell him I'll give you some Papal dispensation for Friday afternoon. But don't screw up. I wanna see them reindeer Sunday. You got it, lieutenant?"

"I got it, Sir," I saluted and left.

On the way back to Company B, I thought about how I was going to make the colonel's reindeer. I needed Ralph and his craft shop to make it all happen.

The colonel wanted holiday reindeer to decorate the ceremonial tank at battalion headquarters.

Ralph was a civilian special services contract worker and the manager of Camp Casey's photography and crafts shop. Ralph was a fun guy and a 30-year-old stray cat from Los Angeles who was billeted with us at the 7th Cav officer's mess. Out of earshot we called him 'Wood Chop'.

While we never heard why the guy left the States for his monastic life in Korea, we assumed it was either girl trouble, wife trouble, or both. He often commented on how he loved his idle time in the sun in the bleachers at Dodger Stadium, so we also assumed he was always at liberty and without a regular daytime job. Korea was just another variety of bleachers for Ralph. It allowed him to pull himself together, get away from his girlfriend or wife for a while, save some money, and see the Far East at government expense.

Ralph was a drinking pal and exceptionally good company. He had creative talent and encouraged my photographic work at his craft shop. I knew I could count on him.

When I got back to Company B, I reported to Capt. Norbert who had already received a call from the colonel.

"Flagg, when are you going to start being a useful officer around here? These colonel boondoggles just take time away from my company and your platoon. I'm your commanding officer, you know."

"Yes, sir," I replied. While he didn't say it, it was clear he was pissed and had finished speaking to me. I saluted and left without a word.

I met Sgt. Temple in the motor pool and told him about my morning, about the colonel's reindeer job, and how upset the captain was.

He almost laughed. "Listen, lieutenant, you gotta start understandin' the chain of command. Your boss is the captain, not the colonel. You daresn't get them confused. Just do whatever the colonel asks you to do, just don't piss off the captain. The right way woulda been to first ask the captain 'politely' for some time off for the colonel's job."

After listening to the sergeant, I got the message, and I especially liked and understood his expression 'daresn't.' *Okay*, I thought, *I'll follow Temple's advice. I'll make the reindeer for the colonel. But I'll 'daresn't' piss off the captain doing it.*

On Friday I was at the company as usual and worked at the motor pool all morning with the platoon. At noon at the officer's mess, I found Ralph Nordick sitting in the lounge before lunch. I asked him if he would help me with the reindeer. I was surprised at his enthusiasm.

"Reindeer? Hell yes! We need some Christmas cheer around here. In the States there'd be parades, shopping, TV specials, and Christmas music. Here, you'd hardly know what year it was. When do we start?"

"I got some time off from the colonel this afternoon. We can start after lunch if you're not too busy."

"Not too busy? The troops don't show up until five. We can start at one. I got some fiberboard in a couple of days ago so we can use it for the reindeer. Have you done any this stuff before?" Ralph asked.

"I did college art for homecomings. I know what to do. The big challenge will be getting the reindeer in the air, like they're flying," I said.

After lunch, like Dorothy and the Tin Man in *The Wizard of Oz*, Ralph and I almost skipped down the road to Nordick's craft shop Quonset in the center of Camp Casey. We were on the colonel's mission to make a pair of reindeer pulling Santa's M-46 in front of the 7th Cav HQ.

Ralph Nordick was the civilian craft shop director at Camp Casey and agreed to use his shop to create Col. Stephenson's reindeer.

"Let's get started," I said, after arriving. "Where's the fiberboard?" Ralph pointed to the 4' x 8' fiberboard stacked in the corner. "Shouldn't we make a template of the reindeer first to transfer to the board?" Ralph suggested.

"No, don't need to," I said. "I can draw directly on the board. Then we'll cut out the drawing directly from the board. Now... if I can only remember how to make a reindeer," and began drawing with a magic marker directly on the board.

"Y'know, you're pretty good," Ralph said, seeing how easily I was visualizing and drawing.

"But be careful Michaelangelo, with that reindeer's long nose, it's beginning to look like a greyhound." he said with a laugh.

After a couple of false starts with nose, neck, and head, I was able to re-draw a nice reindeer on the 4' x 8' sheet. "What do you think," I asked, somewhat pleased with my work.

"I think his nose should still be shorter and cuter. And his back legs should be more flexed - like he's really leaping," Ralph suggested.

We were both becoming flying reindeer experts.

After we settled on some additional anatomical corrections, I began to cut the reindeer out, releasing it from its confinement.

Ralph's shop was well equipped for most Army hobbyists, but had a serious lack of cutting tools for the fiberboard. He had only two saws and neither were electric. One was a compass saw (a narrow-bladed saw for shaping holes), and the other was a coping saw (a thin, stretched blade on an eight-inch square frame). Neither were ideal for cutting fiberboard. With a lot of complaints about Ralph's lack of equipment, I was successful in cutting the silhouettes of the reindeer.

"Ta-da!" I said holding up my cutout after an hour. The only casualty was a broken leg, easily mended with a splint and glue. I also drew and cut out a rack of antlers to insert across the head of my reindeer.

Reindeer-two was easier. We simply traced my cutout onto another fiberboard sheet. Cutting was also easier after my hour's practice. Next came painting and Ralph had everything I needed.

"I don't want to hear any complaints about the paint like my lack of saws, buddy," Ralph said. "We have a complete supply - oils, caseins, watercolors, poster paint. What will it be, sir," as if he was taking a drink order.

The painting went quickly and by four o'clock we had two elegantly rendered reindeer with light brown upper bodies, white faces with big black eyes, white lower bodies, light brown and white legs, orange antlers, and bright red noses. They were ready to fly.

Because the fiberboard was only a quarter-inch thick, we had to glue a couple of wood furring-strips across the inside of each animal's body to provide rigidity for the outdoor wind and weather.

One last step remained: we needed a 10-foot fence post (identical to my fence project) for our flying reindeer, but that could wait until Saturday when I could get my squad to do the fence post work.

It was dark when we finished at four thirty. We congratulated ourselves and promised a drink together at the officer's mess when Ralph finished at his shop.

I walked up the hill to the officer's mess and immediately found Capt. Norbert. He was in the lounge having a drink and reading *Stars and Stripes*. I was mindful of what Sgt. Temple had suggested - asking the captain 'politely' for an hour Saturday to put up the colonel's reindeer.

"Captain Norbert. Thank you very much for giving me time this afternoon for the colonel's job. Ralph Nordeck and I were busy at his shop all afternoon with the colonel's reindeer. All we'll need now is an hour tomorrow to attach them to a fence post in front of HQ. It'll go quickly if I can take a couple of men to set the fence post. I was hoping you'd let me have some time tomorrow morning, sir."

Capt. Norbert swirled his whiskey sour and looked at his glass.

"Okay, I get it," the captain said reluctantly. "I won't stand in the way of the colonel's job. Do what you have to do and get on with it. But don't spend all morning."

"Thank you, sir. I know the colonel will be pleased," I said.

That was my lesson, thanks to Sgt. Temple and his take on the chain of command. I had to make a formal request 'politely,' to avoid having the captain pissed at me. At the same time, the captain reluctantly had to agree to let me finish the colonel's job, to avoid having the colonel pissed at him. The informal transaction respected rank and avoided having anyone pissed at anyone.

Saturday morning went perfectly. In an hour with Nunez's squad, a post hole was dug, and the reindeer were set eight feet high and 30 feet in front of the tank. The painted Santa animals were set four feet apart with pre-cut wooden braces and affixed to the top to the fence post. Phone wires were attached to the tank and the reindeer to look like Santa's reins.

The barren hills and valleys of Camp Casey in December were snow-covered with few available Christmas trees. These enterprising soldiers found one to decorate.

The colonel's Christmas creation was done. Before lunch Saturday, the colonel came by and said, "Those reindeer aren't looking too bad, Flagg" and offered, "Have a drink on me. By the way, where'd you learn to make them reindeer?"

Our unit's tanks, APCs and jeeps operated without heaters, because the Army believed cold running vehicles encourage combat readiness. In subzero temperatures, all vehicles were challenged both day and night. Vehicles not kept running overnight had dead batteries in the morning.

6

Winter Maneuvers

"Make sure your troops are bundled up tomorrow. It's going to be freezing in Kaesong-Munsan, and we'll be camping out there for four days," Capt. Norbert said, concluding his morning briefing. It was Sunday, January 5th and the Korean winter had already set in with snow and freezing weather.

I had been in Company B for three months, and thanks to Sgt. Temple, I was becoming an okay platoon leader. Temple, my platoon sergeant, was always my mentor for 'top-down orders' (from the captain) and 'bottom down orders' (to my platoon's NCOs and privates). Temple had fought in the Korean winter, and I was counting on him to guide me through the winter maneuvers at Kaesong-Munsan, an area about 30 miles west of Camp Casey, near the DMZ. Kaesong-Munsan was going to be a challenge for both men and equipment with temperatures forecast to be zero or below for the coming week.

Our tanks, APCs, and jeeps operated without crew heaters. The Army believed cold running vehicles encouraged alertness for a combat ready crew. But riding in those icy vehicles, combined with the freezing wind from Siberia, extracted every bit of warmth from our parkas, thermal boots, and fur lined caps. Frost bite was common with our troops. Our maneuvers were on the same frozen terrain where the Korean War was fought five years before, but the enemy now wasn't North Korea, it was frostbite.

Sgt. Temple said he was worried about frostbite and suggested a bunk top inspection in the barracks that Sunday to check winter gear. He wanted to make sure all the GIs and Katusas had their issued winter parkas, fur-lined hats, Mickey Mouse boots (thermal boots) and sleeping bags for the Kaesong-Munsan trip.

Temple was especially concerned about some of the Katusas, who had been selling their issued military gear in the village.

Because of our extra preparations for Monday morning, both the sergeant and I were busy Sunday afternoon.

"How's it going, Sgt. Temple?" I said, catching up with him at 4 in the motor pool.

"Our vehicles are all running and topped off (with fuel), sir. The troop's winter gear and sleeping bags check out. The Korean's have all their gear too. If we don't have any breakdowns on the road, we'll be good to go for Kaesong-Munsan."

"Looks like we're ready. Were you in Kaesong during the war?"

"Nope. Our fighting was in the Cheorwon Valley up north. But the Korean winters were the same as they are now. Our tanks were ice boxes. We'll have to watch the troops to prevent frostbite."

With that thought and after exchanging salutes, I wished Temple a good night and caught a ride back to the officer's mess with the captain.

At eight the following morning the entire 7th Cav was on the road with over 100 vehicles and 500 men moving west for Kaesong-Munsan.

We would be joining 8th Army infantry, artillery, and helicopter units in Kaesong-Munsan for mock battles to simulate a North Korean invasion from the DMZ. The Kaesong-Munsan corridor was one of the main invasion routes the North Korean Army used in 1950 to start the war.

Heading out of Camp Casey, my command jeep was leading the 2nd platoon. Sgt. McPherson was riding with me as my radio operator. The sergeant was an outstanding tank commander who normally would have been with his tank, but I needed him that first morning to assist with my command radio. Like Sgt. Temple, McPherson was a career soldier looking forward to his 20 or 30-year retirement after active duty.

"Glad Captain Townsend isn't stopping us today sir," Sgt. McPherson shouted, riding in the back seat by the command radio. "Yep," I shouted, "Townsend is riding with the colonel's column and not worried about our vehicle intervals today. Sergeant, please check the vehicle radios. Make sure they're all working," I shouted over the noise of our jeep hitting every frozen rut.

"All the radios check out except one." McPherson shouted after his radio check.

"Don't tell me..." I shouted, "Nunez?"

"You got it lieutenant - Nunez," McPherson shouted.

"Chou, slow down until Sgt. Nunez's APC passes. We've got to signal him to get his radio back on," I shouted. "Y'sir," Chou shouted back.

I waved all the vehicles by to wait for Nunez's APC. When he passed, we signaled him to get his radio on. He waved back that his radio wasn't working. Once again, we'd have to fix his radio at our first stop. Nunez was Nunez.

After three hours on the road with several stops, including one for Nunez, we reached Kaesong-Munsan. The other 8th Army units were already there for afternoon maneuvers. The 7th Cav's tracked vehicles were formidable fighting machines on open terrain, where speed and maneuverability counted. But on narrow Korean roads in the steep valleys, our tanks were always vulnerable. For that reason, tanks in Korea were often used only in defensive positions, dug into hillside defilades (dug trenches to shield the hull of the tank) with only the cannon and turret exposed.

In contrast to our armored units, the infantry roamed the high ground over rocky and mountainous terrain. Our Kaesong-Munsan mock battles were mainly determined by the foot soldiers who secured the high ground as they had during the Korean War.

With the 8th Army monitors controlling the mock battles, our 7th Cav road-bound vehicle units were most often losing, while the 8th Army infantry on the high ground was usually winning.

Army helicopters also played a role in our winter maneuvers in Korea in 1958, carried troops from one attack position to another - a new tactic then for ground warfare.

Seven years later, helicopters became the mainstay of the U.S. Army battling the PAVN (People's Army of North Vietnam) with the Air Cav moving men and equipment over huge distances to fight the PAVN. Helicopters replaced tanks in the cavalry units in Vietnam. For instance, in the 1965 Battle of La Trang in Vietnam's Central Highlands, the 7th Cav fought a larger PAVN regiment using helicopter troops.

According to *Wikipedia*, "La Trang became the blueprint for the Vietnam War with the Americans relying on air mobility, artillery fire and close air support, while the PAVN neutralized that firepower by quickly engaging American forces at very close range."

Winter maneuvers in peacetime Korea were not the pitched battles of Vietnam, but the tactic of how to deploy infantry, helicopters and armored equipment began to evolve.

Capt. Norbert was on his radio all afternoon giving orders and sending his three platoons over the frozen Kaeson-Munsan terrain.

"2nd platoon, move to checkpoint green and wait. 3rd platoon move to checkpoint yellow and advance a mile south. 1st platoon move to checkpoint blue and take up defensive positions."

At the same time, the 8th Army monitors countermanded the captain's orders all afternoon, moving our units back for their master plan. The captain and the monitors were seldom in agreement. Our platoons were simply pawns moving all afternoon on a giant chess board around Kaeson-Munsan.

At 5 pm, maneuvers were halted for the day and B Company assembled in a bivouac area for the night by a frozen river near Munsan. It was already dark and getting colder.

Mess Sgt. Demsey and his chow truck and crew were there with hot chicken and vegetables, coffee, and dessert on cold metal plates. The mess crew had built a huge bonfire by the chow line to provide added light and heat. It was five degrees above and getting colder and going to be a below-zero night.

"Flagg, Clarkson, Fitzgerald, let's get started," the captain began with his briefing as we were eating. "The fuel truck is late. The driver lost his way. When he gets here, you'll need to refuel all your vehicles tonight. It'll be dark and pumping fuel from 55-gallon drums on the truck is tricky. Make sure your men stay awake until they finish fueling. And no smoking by the fuel truck.

"Set your guard posts and rotate shifts with three hours on, and no cooping (sleeping on guard). You lieutenants should check your guard posts two or three times tonight. We don't want HQ coming around and finding our guards asleep.

"Make sure your Kaesong-Munsan maps are marked for tomorrow with all your check points. Some of you missed your locations today. The monitors didn't help by countermanding my orders today, so tomorrow you've got to let me know when the monitors move you.

"Finally, let the First Sergeant know where you are sleeping tonight. We'll need to find you in case of any emergencies. You got all that?"

"Yes sir." The three lieutenants replied in unison.

I spoke to Temple immediately after the captain's meeting. "Sergeant, set the guard posts tonight with the other platoons. Three hours on and rotate the guards."

"Already done sir. What about refueling? Where's the truck?"

"Don't know, sergeant. Captain said the driver lost his way, and he'll be late. But we got to fuel tonight. Make sure at least one man stays with his vehicle to wait for the fuel truck," I said. "The others can catch their zees if they are not on guard duty."

A half hour later, the HQ fuel truck rolled into the bivouac. All 28 of the company's vehicles were in a line for fuel. It was dark and minus 5 below. Each vehicle had to drive next to the two and a half ton open truck loaded with 55-gallon drums of fuel. A hand-operated rotary pump was inserted into the top of each loaded drum, and a fuel hose then connected to each vehicle's fuel port.

It was a slow process and an ever slower process with cold gloved hands on metal pumps, metal drums and metal fuel ports. With each vehicle there was a risk of spills or fire, but fortunately all 28 vehicles were finally refueled with no problems,t he last one at midnight.

While the crews were in good spirits during the re-fueling, it was a long night before they could crawl into their sleeping bags after midnight.

"Why does it always come down to refueling in the dark?" Sgt. McPherson complained while waiting by his tank. McPherson was always cheerful, but the cold and dark night even got to him.

"Sorry, sergeant," I replied, "The driver lost his way. Shit happens."

After the last vehicle was refueled, I checked my guards to make sure they were awake. It had been an eighteen-hour day with adrenalin rushes and boring waits. It was a day with tanks stuck in frozen rice paddies and a flurry of radio orders, wrong map coordinates, missing check points and 8th Army monitors moving units all around.

Finally, at 1 am, it was my turn in my sleeping bag. But that too was a challenge, crawling into a frozen bag that had been sitting in a cold jeep all day. To conserve body heat, I kept all my clothes on. I took my boots off, but it was still a shivering 15 minutes before I regained enough body heat to raise the sleeping bag temperature above zero. After three hours of this cold and uncomfortable rest, I checked the guards again and was awake at 5 am to start my day.

Sgt. Dempsey already had his chow line going and a bonfire blazing at 5. It was below zero and dark, but I saw Sgt. Temple standing close to the fire, indifferent to the cold, in his shirtsleeves and red suspenders, rousting the troops for their breakfast. As always, he led by example.

"Wake up you guys. It's time for chow. Don't let a little cold weather keep you from a hot breakfast."

Following Temple's exhortations, the troops roused and fell in line for Sgt. Dempsey's scrambled eggs, bacon, and hot coffee. Day two was underway.

The second day was like the first. The infantry was on the hills. The tanks, APCs and jeeps were in the valleys. The captain was endlessly on the radio moving his platoons from one check point to another, while the 8th Army monitors was moving them back.

Our mess sergeant was active all four days of maneuvers in providing hot breakfasts, lunches, and dinners.

With the temperature hovering at zero, bright sunshine felt great on our faces while we waited for our next orders with our cold vehicles standing still. At noon, we paused to warm ourselves with Sgt. Dempsey's bonfire and hot chili.

A soldier named Gonzales yelled to Dempsey, "Hey Sarge. You forgot the jalapenos in the chili. We need mas caliente if we're gonna keep goin' through this cold."

Dempsey's chow line was the singular best event that day while we played mock warfare with the captain and the monitors.

At 5 the winter maneuvers were halted for the second day. Company B bivouacked in a new area by a frozen canal. This time, the HQ fuel truck was waiting and refueling began immediately. Sgt. Dempsey's chow line and bonfire provided a warm welcome. By the second day, the vehicle crews were becoming accustomed to the Siberian blasts. But preventing frostbite to fingers and toes was becoming a 24-hour challenge. I experienced numbness on my toes while riding in the cold jeep and sleeping in the cold sleeping bag. I was not alone. Several troops were frostbitten during our four-day outing. I still have numbness on my feet during cold weather, the result of that Korean frostbite winter.

The entire 7th Cav battalion was on the road with 500 men, nine M-48 tanks, nine APCs, 51 jeeps, and supply/mess trucks in the freezing cold to winter maneuvers at Kaesong-Munsan.

Another reminder of the icy temperature was on our bare skin while relieving ourselves. There were no latrines in Kaesong-Munsan, only dug slit trenches for urinating and defecating. With pants down, the arctic blast was forever memorable.

The outdoor conditions also created a problem for washing and shaving. At 5 in the morning, without hot water, mirrors, or lighting, the troops were indifferent to washing and shaving. One GI managed to shave only one side of his face. Another only shaved his chin. A third did not wash or shave and had a beard-shaped mask that outlined his eyes and mouth with road dust. Both Sgt. Temple and I were aware that our troops were unshaven and unwashed. I asked him what to do.

"Those horses ain't goin' anywhere near the frozen' water today, sir," the sergeant said. "Best, we let 'em be. We'll clean 'em up tomorrow. We got bigger things to worry about today," was Sgt. Temple's focus on the freezing weather and the maneuvers at hand.

In the subzero temperature, vehicles that were not running throughout the night had dead batteries in the morning. Jumper cables became the only alternative to firing up dead engines. Jeeps with wheels frozen to the ground had to be hauled out by tanks. And it finally became a necessity to keep our vehicles running all night so they were ready to go in the morning. The Korean winter and frigid nights took their toll on vehicles.

The winters also took their toll on the local Koreans who were without electricity, running water or sewage. Mud brick buildings with thatched roofs were poorly insulated, and firewood was the only fuel for heat. We saw momma-sans with their kids, washing clothes in the local icy streams, and papa-sans collecting firewood, while tending to their oxen and vegetable patches. Those vegetable patches produced ingredients for an essential local staple - kimchi. Kimchi had a powerful aroma - a mix of cabbage, chilies, radishes, herbs, and fermented fish stock. We often knew where villages were, even a mile away, simply by the smell of kimchi in the air.

By the third and fourth day, we knew what to expect from our winter maneuvers and the temperature. The captain and the monitors were entirely in charge - we were the pawns in their game that stretched over the entire 15-mile area of Kaesong-Munsan.

During a lull, Sgt. Temple mentioned his experience in Korea five years before. "Sir, things ain't changed much since we were fightin' up in Cheorwon. The infantry is still a-holdin' the high ground, and tanks are still sittin' in the valley. It ain't a great strategy to have the enemy shootin' down on you."

Winter maneuvers was simply the Army's test to see if we could survive the cold and still fight. Sgt. McPherson was riding with me going back to Camp Casey on Thursday, and as we chatted over the road noise. I asked him what he thought of our winter maneuvers.

"Lieutenant, I just spent four days runnin' around in the friggin' cold in Kaesong-Munsan and almost got frostbite. I don't need any more practice to be miserable."

Col. Edmundson loved parties. He sent me to Seoul to pick-up of a fabulous 5-string banjo player who played at the Grand Ole Opry and was stationed in Seoul as a US Army dentist. This was the start of great Saturday night parties at the officers' mess. Col. Edmundson and Capt. Tiaphin brought down the house with 'The Wabash Cannonball".

7

Saturday Night at the Officer's Mess

Colonel Edmundson loved parties.

After arriving in March as commanding officer of the 7th Cav, he began a Saturday night ritual at the officer's mess of a party or a musical evening to liven up the weekend.

He was an accomplished blue grass guitarist, knew dozens of country songs, and was good at singing and playing. With his arrival, we could look forward to entertaining Saturday nights after an active week in camp or in the field.

"Flagg, you know how to get to Seoul and the 8th Army BOQs, don't you?" he asked me one Friday at lunch.

"Yes sir," I answered.

"Well, I got a job for you and your driver tomorrow morning," he said.

Colonel Edmundson and I had become poker buddies, even though our ranks were vastly different. Our friendship started during our Saturday and Sunday afternoon poker games and led to special assignments he found for me during regular duty days.

"There's a fabulous banjo player who has just reported for duty in Seoul, and I have invited him to come up this weekend to play for us. His name is Capt. James Tiaphin, and he's a dentist working at the 8th Army medical center in Seoul. He's played at the Grand Ole Opry, and he's agreed to come up with his five-string banjo to play for us tomorrow night.

"I'll call Capt. Ramsey and tell him I need you for an important job tomorrow. I'll give you the details at dinner tonight. I want to make sure Capt. Tiaphin gets the VIP treatment. You got it, Flagg?

Have your driver ready first thing tomorrow morning at seven."

"Sir." I understood my orders.

I spoke to Capt. Ramsey, my new company commander, after lunch, and he said he just got the call from the colonel, and rolled his eyes, "Flagg, what do you got going with the colonel? Are you his valet or something?"

"Don't know, sir. Guess there's no one at headquarters for special jobs."

At Friday night dinner, the colonel gave me the address of Tiaphin's BOQ in the 8th Army compound. He said he also wanted me to go to the 8th Army PX first to shop for a special dinner item for Saturday night.

"Flagg, the Camp Casey PX here doesn't have zilch, but the 8th Army PX is loaded. I want you to buy 10 packages of frozen oysters for Saturday night's hors d'oeuvres. Have S-1 reimburse you when you get the receipt.

The colonel gave me the order to buy frozen oysters at Seoul's Army PX (post exchange) to serve as special hors d'oeuvres.

"We don't want Tiaphin thinking we're not sophisticated just because we're up north here next to the DMZ. We want to do it up right for our guest," he said with his usual chuckle.

The following morning at seven Chou had my jeep ready at the officer's mess with its canvas top up. We were off to Seoul in the rain for the colonel's two missions: chauffeuring a five-string banjo player back from Seoul; and serving as the oyster officer for the officer's mess.

After a rainy two hours on Route One to Seoul, we arrived at the 8th Army compound and immediately went to the PX. Capt. Tiaphin would not be ready until Noon.

The colonel was right about the PX. The 8th Army base PX had everything and was like a shopping mall back in the States. Seoul had Army-dependent families and U.S. government employees stationed there, and the PX was their main source for American food and shopping. At the frozen food section, I found only eight packages of frozen oysters and took all of them. I also picked up two packages of frozen fish sticks, so the colonel wouldn't think I'd ignored his specific orders to get 10 oysters.

After the oysters and fish sticks, I took a walk around the PX to see what else it had. The PX in Seoul was the same size as the main PX at Fort Knox. It had everything, from Wheaties to Rolex watches to dress-blue uniforms. After I had my walk-around, I went to the checkout and paid for the frozen food, making sure I kept the receipt. It was still early for Tiaphin, so I had a coffee at the snack bar and bought one for Chou who was waiting in the rain, parked outside the PX.

The next stop was Tiaphin's BOQ but we were still early. The captain had his dental duty until noon. At 12:15 we saw a captain in a raincoat walking toward us, and we assumed it was Tiaphin. He saw us waiting and waved.

He had a wonderful, toothy smile and introduced himself with no pretensions. "Hi. I'm Jimmy and I gotta go to my BOQ first to wash up and get my things. How y'all doing?"

Though "Jimmy" presented himself as a guy from Kentucky working for the Army in Seoul as a dentist, he was in fact an officer with captain's bars on his uniform.

I said, "No prob Captain. I'm Lt. Flagg. Happy to meet you. Colonel Edmundson has told me a lot about you. We're here to give you a VIP ride up to 7th Cav."

"I'm impressed that Col. Edmundson found me and knew 'bout my playin' at the 'Opry.' Looking forward to meetin' him in person. Must be quite a guy."

I assured him he was.

After going to his quarters and bringing his banjo and his weekend bag, we were ready to go. I asked him if he wanted to stop for lunch at the officer's mess, and he said, "No thanks, lieutenant, let's just go. How long's the trip?"

"Maybe two hours. It's only 40 miles, but it's raining, and my driver Chou has to drive between the puddles," I answered.

Chou started weaving his way out of the 8th Army compound, onto Seoul traffic, and finally onto our unpaved Route One heading north.

"I've been on my share of country roads, and as long as this buggy has springs, I'm good to go," Jimmy said. "You say your name is Chou? How y'all doin' Chou? Jimmy here," Jimmie said, introducing himself to Chou from the back seat and patting him on the shoulder. Captain 'Jimmie' was going to be a total delight.

We arrived at the 7th Cav officer's mess about 3:30 as the colonel was finishing his afternoon poker game. With a huge smile the colonel greeted Capt. Tiaphin. "Welcome to the 7th Cav, captain. We have heard a lot about you. Glad you could come up for a visit. Can't wait to hear your pickin'."

"Pleasure's mine colonel. Don't know how you found me, but anytime I can play, I'm game," Capt. Jimmy replied with his big toothy smile. It was clear they were both country, the captain from eastern Kentucky, the colonel from western Virginia.

After introductions to the officers at the colonel's poker table, and the other officers at the bar, the colonel invited the captain to freshen up and drop his gear at a spare bedroom at the captain's BOQ. "See you back here at the bar around 5 for drinks. We have a special dinner planned," the colonel said with glee, "bring your appetite and your Grand Ole Opry banjo."

After drinks at 5 and the colonel's dinner at 6 with special hors d'oeuvres of oysters Rockefeller - the best our Korean cook could do was to broil the oysters without shells and put them on small plates.

The entree was steak and wine. Then the 35 officers adjourned to the bar for some serious drinking and the colonel's country music fest.

The star of the show was Captain Jimmy and his five-string banjo. He was simply dazzling with his instrument. He played standing up, sitting down, and holding the banjo behind his head. He even played blindfolded. His country music filled the room, with his right thumb's metal pick plucking the fifth string while his right index and middle fingers strummed the strings – all done with such speed and dexterity there was only a blur of his right hand on the strings and his left on the finger board.

Five-string banjos have a deep resonant twang achieved by picking the strings closer to the bridge. Tiaphin's virtuosity was full of that twang, and he brought down the house picking 'Wabash Cannonball' with the colonel going with him on his blue grass guitar.

Bringing the Grand Ole Opry to the 7th Cav that Saturday night was Colonel Edmundson's special treat, and Captain Jimmy returned many times. It was the colonel's love of Blue Grass and Capt. Jimmy's love of playing that made those Saturday night's so special.

Driving Capt. Jimmie back to Seoul that Sunday afternoon, I told him how much in awe I was of his playing.

"Heck, that's ain't nothin, man.' You should hear those ole boys back in the Kentucky hills," he said. "They really know how to play a three fingered claw. They started me at five and I've been doin' it ever since."

The USO troupe's performances always erupted in standing ovations after each camp show. The GIs loved the American women and their talent. The audience was on their feet at the end of each performance, clapping, dancing, laughing, being totally taken by the troupe's amazing talent.

As the escort officer for the USO troupe, my job was to organize the ground transportation from camp to camp and be in charge of camp liaison to arrange sleeping accommodations, meals, rehearsals, and show times. The troupe and I rode in a deuce-and-a-half truck the entire 240 miles from Busan to Camp Casey to entertain at 11 Army camps down and up South Korea.

8

My USO Tour

"When you're finished lunch, report to my office, I have a job for you," Col. Edmundson said, on a June day as he passed by my table on the way out of the officer's mess.

Col. Edmundson and I hit it off the first week he took over the command of the 7th Cav back in March. We both loved poker. Every Saturday and Sunday afternoon we played poker together at the officer's mess with the other officers. The colonel was always in charge and appointed me his shill.

Col. Edmundson and I bonded, because we both loved poker. Our poker table rapport helped when he chose me to escort a USO troupe all over South Korea.

"Flagg, get our poker game together this afternoon around 1:30. Round up the usual suspects and make sure we have seven for the game."

"Sir," and I was on my task to coax five other players to join the colonel and me at poker.

Some officers were good players and wanted in. Others were more reluctant, had lost money and opted out. For those who played, it was a spirited game with our colonel showing an entertaining side of his personality.

Col Edmundson was a river boat gambler at heart, and afternoons were a pure delight with his jokes, stories, and friendly ribbing. He was an entertainer and an excellent blue grass guitarist, playing and singing country tunes like 'The Wabash Cannonball.' I especially enjoyed playing poker with this older man whom I truly admired. He liked me too, although our difference in rank was evident. He had no inhibitions of reminding me and all the lieutenants that, "Lieutenants are lower than whale shit, and worthless as teats on a boar hog." The Colonel's flair for profanity was reminiscent of General Patton and his barracks room language of WW II.

As the escort officer for the USO troupe, my first job was to meet my six USO performers in Buson, the port city on the southeast coast of South Korea.

At lunch, I told Capt. Ramsey that the colonel wanted to see me, and after lunch, he dropped me off at the battalion HQ, where I reported to the colonel.

"Sir, reporting as ordered," I saluted.

The colonel returned the salute, and asked, "Flagg, how long have you been here?"

"Sir. I came last October. It's going on 8 months."

"Well, I guess you've been here long enough to know your way around. You need a break from the poker table," he said with a serious look.

"This morning we got a telex from 8th Army. They need an escort officer for a USO tour coming to entertain the troops around Korea for the next two weeks." He paused theatrically, like he was about to lay down a full house. "I have to assign an officer for the job that I'd really like to do myself." Another pause.

"Flagg, I should have my head examined, but I'm assigning you as escort officer." he said with a huge grin.

"Now check with Capt. Ramsey before you go and pack your bags. You'll have to report to Special Services, 8th Army in Seoul tomorrow morning and they'll tell you what to do. Take a jeep at seven. We'll clear everything with your captain. One word: don't fuck up! Now get the hell out of here. I got some real work to do. Dismissed."

"Sir." I saluted and practically danced out of his office and hitched a ride to Company B to tell the captain.

I met Capt. Ramsey in at the company HQ and told him about the USO tour that Col. Edmundson had assigned to me.

Capt. Ramsey was not at all pleased, and almost hostile. "Why the hell's Edmundson yanking my officer for a God damn USO boondoggle?

"How'd he pick you? You knew we got live fire next week. We need every officer on the tank firing range for safety reasons alone. How long will you be gone?"

"Sir, I'm not sure, but I think it'll be about two weeks," I said softly.

"That is simply unacceptable, lieutenant! I suggest you go back and tell Col. Edmundson to pick someone else." he said, fuming and puffing like a threatened frog.

I answered carefully, "Sir, I honestly don't know why he picked me. I'm terribly sorry I'll miss live fire. But I got my orders directly from the colonel ..."

I knew then that I would be on Capt. Ramsey's permanent shit list. The captain was a nuclear physicist and was at the Army's atomic bomb testing site in Nevada in 1957. He was a scientist who went by the book. But Ramsey didn't own me and the Army did. I was going to follow Colonel Edmundson's orders.

Later in the day at the motor pool, I met Master Sgt. White, my new platoon sergeant, who had replaced Master Sgt. Temple in May.

I thought Temple was irreplaceable because he was such an outstanding person and great soldier. Sgt. Temple had been a major who'd fought in the Korean War five years before. He'd been my mentor and counselor for the past eight months and taught me to lead men by example.

But I was wrong: no one in the Army is irreplaceable.

Sgt. White was an entirely different kind of NCO than Temple but just as effective. He was black, 6 '2", weighed 220 lbs. and had a jovial presence. He commanded in a soft-spoken voice. I was lucky to have two of the best mentors a young lieutenant could have.

I told Sgt. White about my USO tour, and he laughed until he almost shook. "Lieutenant, you best stay away from them dancing girls: they'll only going to get you in a heap of trouble. We want to see you back here in one piece."

He nodded when I told him about the captain being angry about my missing live fire. "Don't you worry about the captain and live fire. The troops know what to do, and I'll be there in case they don't. You go and have a ball- the Army's paying for it," he said with another big laugh.

The next morning Chou picked me up at 7, and we drove off to Seoul and the 8th Army Special Services office.

The Republic of Korea was constructing new facilities in Seoul back then, but the U.S. Army had become the major builder in Seoul, with offices, motor pools, barracks, and every kind of military facility, from engineering depots to football fields. The Army was also in charge of re-building Korean roads, bridges, tunnels, and railroads – all critical wartime transportation routes. The City of Seoul had its own building program that included food markets, retail shops, and every type of residential housing in a great sprawl for the over 2 million inhabitants then. Today, Seoul's population is 10 million and is a world class metropolitan city.

After my morning briefing at Special Services, they gave me written orders, the USO's itinerary, a map, and the contact names of special services officers at every camp the USO tour would visit.

The USO troop would be arriving by overnight ferry from Japan, and landing the following morning in Busan, on the southeastern tip of Korea.

I was to meet them on the dock in Busan at 9 am, and Special Services had arranged for an overnight train to get me there. At 8 pm I boarded a sleeper car for Busan at Seoul's central station.

The train consisted of ten ancient Japanese Wagon-Lit cars the Army had refurbished when it re-laid the narrow-gauge track that had been torn up during the war. I had my own private compartment with a pull-down bed for the eight-hour trip (a 210-mile journey that was slow and rickety at 30 miles an hour). I had plenty of time to study the map, the itinerary, and the contact names at each camp. The last camp on the list was my own Camp Casey.

My written orders read that I was in charge of arranging ground transportation from base to base for the USO troupe and their gear, which included sound and lighting equipment, musical instruments, wardrobes, and personal luggage.

I was also in charge of base liaison to arrange sleeping accommodations, meals, rehearsals and show times. And I would be the 'go-to-guy' in any emergencies.

It looked like a daunting job, but it was in fact easy. What made it easy was the secret ingredient - the "round eyes" (American women entertainers). Every camp had GIs wanting to see live entertainment, and American women entertainers were especially welcomed. That translated to overwhelming camp hospitality from the commanding general down to the PFCs unloading trucks. They all wanted their stateside entertainers (especially the round eyes) to feel welcome and comfortable at their camp. That included providing abundant drinks and food, comfortable sleeping accommodations, transportation, and adequate downtime at the officer's mess. The hospitality paid off in great shows from the appreciative performers.

The USO's first show was at Camp Hialeah in Busan the night we arrived. Hialeah set the tone for every camp we visited over the 240-miles of the South Korean peninsula.

The moment my train arrived in Busan, I phoned the special services officer listed on my contact sheet for Camp Hialeah, our first tour stop. Capt. Jacobs was the officer in charge and had received a telex from 8th Army with information that the USO troupe would be arriving at his camp that day.

"Good morning, captain," I said phoning him. "I'm Lt. Flagg, the escort officer for the USO troupe coming to Camp Hialeah for tonight's show. The troupe will be landing by ferry from Japan about 9 this morning, and I'm hoping you can send a deuce-and-a-half (a two and half ton truck with a canvas top and benches) to meet us at Dock B1 at 9 to pick us up. If you can, please send another GI along to assist the women getting into the truck. Would appreciate it."

"No prob,' lieutenant. We knew you're coming and gotcha' covered," Jacobs said. "See you in camp."

After my call, I got a ride to the pier to meet the ferry. The itinerary said the USO troupe was arriving at the ferry Dock B1 at 9. When the ship landed, I had no trouble spotting the six American entertainers. They were easily distinguishable from the rag tag GIs, Japanese sailors and Korean nationals also debarking the ship. I waved to the troupe and walked over to meet them by the gangplank as they waited for their luggage and show cases. "Hi. I'm Lt. Flagg, and I'm your escort officer. Welcome to Korea," I said with a big smile, sizing up the troupe.

There were three very good-looking women, two taller ladies and a shorter one with a great body. They were in their twenties and wearing leisurewear for their overnight trip. What struck me most was how poised and nonchalant they seemed.

There were also three men, two younger guys and an older man who seemed in charge. "Hi, I'm Jake," he said introducing himself. "We're still waiting for our luggage and show cases. We didn't get much sleep last night with the rough seas. Hope you can help us out with a shower and some rest. And come meet Keith Jordan and Larry Roberts, our musicians."

I shook hands and nodded hello.

"The girls over there are Patty Smith, Maureen O'Shea, and Barbara Gordon," said Jake. "You're not here by yourself, are you?"

"Not to worry. It's Jake, right?" I reassured him with a little laugh. "The truck will be here a few minutes. We gotcha covered. And I am sorry about your boat trip. We'll get you some rest and a shower when we get to camp."

I walked over to the women and introduced myself while the ship's crew brought down their luggage and show cases. I was beginning to worry that the truck might be delayed.

To my surprise, when the troops arrived a few minutes later, there was not one but two trucks and a jeepload of GIs. The camp went all out with plenty of help for our entertainers with their luggage and show cases. The jeepload of GI's couldn't do enough for us. They insisted on carrying everyone's personal luggage and loading them with the troupe's 10 fiberboard cases in the first truck.

The GIs helped everybody get into the second truck with the open tailgate, providing hand-holding, arm-holding, and waist-holding help for the ladies. Two GIs rode with us next to the closed tail gate to ensure that everyone was seated and safe during the bumpy ride to camp.

When we arrived at Camp Hialeah, we were driven to our guest quarters (a Quonset that was formerly a BOQ and was now nicely appointed with single rooms and baths). I told the driver of the first truck to drop off the troupe's personal luggage at the guest quarters and deliver the ten show cases to the camp's theater for the afternoon rehearsal.

At the guest quarters, the USO troupe was greeted by the camp adjutant, a Lt. colonel who welcomed us and apologized that the commanding general couldn't be there personally. As a second lieutenant, I was experiencing hospitality I had never seen before. What was amazing what a few round eyes could achieve in a women-starved GI camp. I was going to enjoy being a USO escort officer.

After settling in their private rooms for a little rest and a shower, the troupe was invited to the officer's mess for a private lunch with the adjutant, his staff, and the special services officer. Wine was served for the guests, something unheard of at most officer's messes during duty hours.

The rehearsal in the afternoon went flawlessly. The troupe had performed dozens of times in Japan and Okinawa and had their show down pat. They had excellent support from the NCO theater manager and his crew who set the stage for the rehearsal. The crew had everything working at their 300-seat theater: the audio, the mikes and recording systems, the spotlights and back lighting, the props, and all staging equipment for the two-hour show.

There was only one problem. There were no private dressing rooms. The players had to do their costuming and make-up for the evening performance in their bedrooms, then ride in open jeeps to the theater.

Those rides created quite a stir for GIs walking in the camp street, as costumed and made-up American women with long legs and knees up sped by in open jeeps. The USO tour was listed in the camp bulletin, but the impromptu knees-up show at 7 created all the buzz needed to fill the room at 8. Over 300 GIs were eagerly waiting when the show opened at Camp Hialeah.

The USO had assembled the troupe in Detroit, and they had been assigned a plum Far East tour with bases in Japan, Okinawa, and Korea. The GIs at each base saw a great show with an extraordinarily talented and surprisingly small troupe of only six performers.

Patty Smith was the first, a perky 25-year-old buxom blonde and a former Michigan high school baton twirling champion who opened the show. She had incredible dexterity, twirling three batons to "Star and Stripes Forever" and other Sousa marches.

Her final show-stopping feat was tossing a baton 20 feet in the air, while pirouetting three times while twirling two other batons before catching the first. She wore a majorette's costume with gold and silver braid across her ample bodice, a short skirt and white boots that accentuated her curvaceous figure.

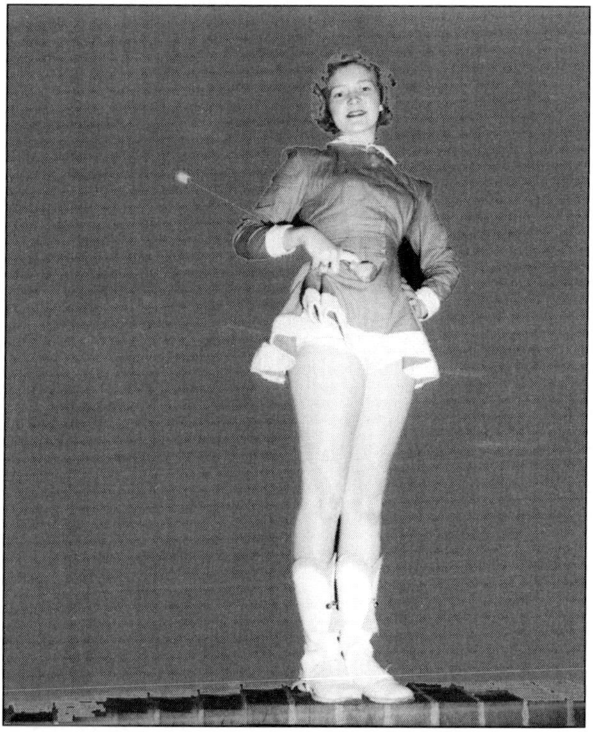

Patty Smith was the show opener and a buxom Michigan high school baton twirling champion with incredible dexterity twirling three batons.

The young GIs loved seeing her extraordinary skill and wholesome manner, reminding them perhaps of their own high school sweethearts or loved ones back home.

Maureen O'Shea followed, a talented 26-year-old soprano with beautiful red hair and an Irish lilt in her voice. She sang Rodgers and Hart and George Gershwin ballads "Thou Swell," "I Got Rhythm," "Our Love Is Here to Stay," belting them with her Broadway flair. She wore a beautiful green short-skirted dress with sequins that emphasized her model's body and beautiful long legs.

Barbara Gordon was the lead singer and a sultry 28-year-old brunette with a sexy alto voice. She wore a stunning full length red silk gown with V neckline and slits to emphasize her graceful walk.

The GI's wolf whistled her as she sauntered slowly to center stage, shyly acknowledged her admirers with a little Marilyn Monroe smile. Her love songs were classic: "C'est Si Bon" (It's so Good), "Let's Do It "and "Santa Baby," Eartha's Kitt's '50s smash hit, even though it wasn't Christmas. The GIs swooned when she sang and applauded and whistled when she finished.

Keith Jordan was one of the talented accompanists on trumpet and rhythm guitar. He was handsome and the most versatile of the musicians. At 28 he was an accomplished studio player at the recording studios in Detroit. He said the USO tour was a Far East fling from his daily routine. He and Patty Smith became a romantic couple during their Korean tour.

Larry Roberts was also a wonderful accompanist and an accomplished bass and keyboard player, filling the theater every night with a full range of orchestral sounds from his single keyboard. He was 30 and played jazz clubs and cocktail lounges all over Detroit.

Jake Nichols was the final act on stage and the star of the show, a 40-year-old comedian (and an unexpectedly fabulous alto sax player) who brought down the house every night. He weighed 230 pounds, was 5 foot 8 and got a laugh by simply walking on stage, much like Jackie Gleason, another comedian with girth.

His opening routine was hilarious, telling one blockbuster joke after another with riotous punchlines. The young GI audience loved his humor and roared with laughter and applause. After the jokes, he walked to the wings and brought out a folding chair and a small toy mule on a rocker. Again, with a deadpan expression (the audience was already beginning to laugh) he sat down and pulled a piece of string from his pocket, tying it to the tip of a long pencil (more laughter).

To everyone's surprise, the PA then blasted Frankie Lane's recording of "Mule Train," with Jake sitting as the mule driver, whipping his tiny toy mule with the string. The audience went wild, roaring with laughter, clapping and cheering Jake's brilliant pantomime.

After Jake's riotous "Mule Train," he had one more surprise. He walked again to the wings and brought out a saxophone, awkwardly toting it to center stage. (More tentative laughter. What's the fat man going to do with that saxophone now? Where's the joke?). Jake brought the mouthpiece to his lips and tooted a few honks (audience again laughed, still expecting another 'Mule Train' pantomime).

Jake Nichols was the final act, and he brought down the house every night with hilarious jokes, a riotous pantomime, and his total (and unexpected) virtuosity on his saxophone.

To their utter surprise, what followed was Jake's fantastic mastery of the sax: he not only knew how to play, he WAILED, opening with "Take Five," the Dave Brubeck and Paul Desmond's classic sax opus. His final act was a showstopper with 15 minutes of high energy jazz, with Keith and Larry as his side men. The GI crowd was on their feet, dancing, applauding, cheering, and rocking to Jake's jazz tempo. The show closed with all the performers returning on stage and rocking to Jake's sax. Tremendous applause followed with a standing ovation.

The show was a smashing success. I was in the audience for all eleven shows on the tour and the audience never failed to rise for standing ovations. The show had it all: beautiful round eyes from home, jokes, humor, love songs, ballads, and a jam session that rocked the house.

Ten more smash hits followed at ten Korean U.S. Army bases. Each was as successful as the first show at Camp Hialeah. Moving up the Korean peninsula, we visited Camp Custer, Camp Carrol, Camp Page, Camp Market, Camp Hovey, Fort Chandler, and the Osan Air Force Base. The troupe traveled over 240 miles of dusty roads and out-of-the-way camps in an Army deuce-and-a-half truck.

Each camp provided the same great hospitality as Camp Hialeah. Yet each camp was different from the size and configuration of the theater to the living space, to the long drives to reach each one. Fortunately, the USOers were troopers who had worked out-of-the-way cocktail lounges and jazz clubs throughout Detroit. Dusty truck rides, uncomfortable beds, no hot water, and missing meals were no big deal, and only seemed to heighten their Korean adventure.

After performing at the Seoul area camps, we trucked north toward the DMZ, with shows at Camp Jackson, Camp Castle, and finally Camp Casey. The USOers were now experienced deuce-and-a-half travelers, with the women wearing GI fatigues and hats given to them by adoring GIs. They also brought cushions and blankets to soften the hard benches for those long rides. Some of the women even talked drivers into letting them ride in the front, providing a more comfortable ride (and a friendlier ride for the drivers).

The troupe finally arrived at Camp Casey, their last stop, and the end of the tour. For me, it was like coming home and bringing the circus. The USOers were formally greeted by the commanding general and taken to their guest quarters to freshen up. After a little rest. I couldn't wait to take them to the 7th Cav officers mess to meet Col. Edmundson and the other officers for lunch.

"Where the hell have you been hiding these beautiful ladies, Flagg?' the colonel said with a big smile, welcoming the troupe and me to the 7th Cav mess. All the officers were there, tumbling over themselves to meet the performers, and especially Patty, Maureen, and Barbara. After introductions and more merriment (but no alcohol until 4 o'clock) Maureen and Barbara sat on either side of the colonel, and Patty, Keith, Larry, and Jake mixed with the captains.

The colonel was in his element, entertaining and making all the performers feel welcome. "We are delighted to have you as our guests and I understand we've arranged tank rides for you this afternoon. After the tank rides, you all will qualify as honorary members of the 7th Cavalry. We have never made this designation before (with a theatrical pause), because we've never had a USO troupe for lunch before (laughter and applause). But we're starting now. Please feel free to come back any time. Enjoy your tank rides this afternoon." (Standing applause from the officers.)

After lunch, and more good humor and goodbyes, our USOers were off in waiting jeeps for the tank rides I promised at Company B. Before leaving, I made sure the troupe met Capt. Ramsey, who was at his charming best as he personally escorted the two ladies to Company B and the waiting tanks.

Sgt. White and all the other NCOs were there to help the troupe with their helmets and climb up and into our 45-ton M48 tanks. Once they were in the turret seats, they were helped again in putting on their intercom headsets and being strapped in. It was an exhilarating and unique experience for my six USO friends, wearing headsets and helmets, and shouting and laughing over the intercom to the driver while they were driven over a bouncy loop around camp.

Everyone raved about their tank rides, and Patty Smith was especially excited and gushed, "I was so scared inside that huge tank. But it's something I'll remember for the rest of my life. And those soldiers were so nice, getting me up and in the tank and helping with my helmet and headset. I hope you'll thank them for me."

I knew how raunchy my guys could be, so I was hoping that was all she remembered about their help!

During the troupe's tank rides, Sgt. White pulled me aside and told me. "Lieutenant, no need to worry 'bout live fire last week. It went off without a hitch and the captain was fine. While we missed you, I was hoping none of them pretty girls were leading you astray," he said with big laugh.

After the tank rides, we jeeped to the camp theater to set up for rehearsal. Afterwards, it was back at the guest quarters for a short rest, and a quick dinner. The troupe had to do costuming and make-up in their bedrooms before their last 'legs and knees up preview' as they were whizzed by the GIs in open jeeps. Camp Casey's theatre had no private dressing rooms!

Each camp has a base theater but they were all quite different. Some had dressing rooms and others required the performers to dress in their bedrooms.

The show was a smash hit for the 11th time, and for me, I knew I brought a great USO show home to Camp Casey. That night I stayed in my own BOQ for the first time in two weeks.

The next morning the troupe and I were on the dusty road back down to Seoul, for the last deuce-and-a-half ride. The tour was over. It was my last official duty to deliver the players back to Special Service and their guest quarters at the 8th Army compound in Seoul. They would be flying out the next day.

When we said our goodbye, I realized how close I had grown to my six amazing new friends in two weeks. I was not only their escort officer, but I was also a participant in their long rides, a different theater every night, a different bed every night, no hot water, missing meals, and roughing it in Korea.

I was also in their audience every night, enjoying their show and participating in the standing ovations with the GIs. The troupe knew I loved being with them and enjoyed the part I played in bringing their show to the camps in Korea. They gave me a jade ring and big hugs and kisses when they said goodbye. I had been part of their family for two weeks in Korea.

Our summer maneuvers had the entire 7th Cav on the road to Chuncheon, a mountainous area east of Camp Casey.

9

Summer Maneuvers

Captain Ramsey called a pre-op meeting to tell us we were going the following Monday for four days of summer maneuvers.

The Army's annual training schedule in Korea always included both a winter and summer field op to test a unit's capability in actual wartime conditions.

"Flagg, I want you to do a map survey of the route we're taking next week. We're going east to somewhere around Chuncheon and I hear it's mountainous and the roads can be tricky. I wanna make sure all our tracked vehicles can make it. Get back to me this afternoon with your report," he said.

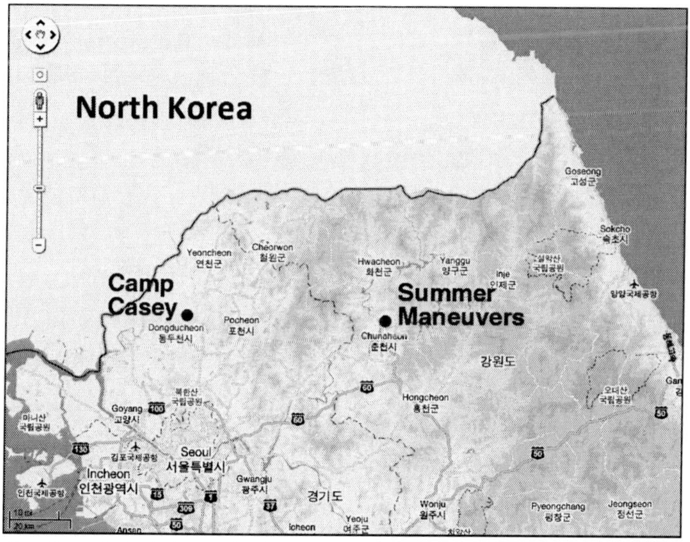

The Army's training schedule always included winter and summer field ops. Our summer maneuvers had the entire 7th Cav on the road to Chuncheon, a mountainous area east of Camp Casey.

This was the start of a long check list with Captain Ramsey continuing to dish orders to me and my fellow lieutenants, Albert Darden, and Tim Humphreys.

Company B, like the other companies in the 7th Cav, had 120 troops at any one time. About 50% were American GIs at various stages of their 13-month tours with constant turnovers of those experienced officers, noncoms, and GIs. The other 50% of the troops were ROK Army supplements.

Captain Ramsey was a scientist at heart, and not necessarily a field op guy, and his by-the-book leadership of our mixed-and-matched troop unit was always problematic. But Ramsey always did his best, even with the limitations of men and equipment. That was why he had called a pre-op meeting that Thursday morning.

"Last item on my list: Make sure your emergency provisions are properly stowed in each vehicle and in good condition," said the captain. "Those C-rations, emergency water, and First Aid kits are to be opened only in an actual enemy invasion. That's an 8th Army directive.

"That's all on my list. Keep your crews working on your vehicles today and tomorrow. Let me know if any of the vehicles are on the disabled list. Questions? Dismissed." The lieutenants rose, saluted, and went to the motor pool.

When I caught up with Sgt. White at the motor pool, he was teaching our new Katusas how to knock mud off the undercarriages of the tanks and APCs with a broomstick. Sgt. White and I were both concerned that our Katusas were just riding in our APCs without learning anything about them. We did what training we could, but it always came back to two basics: keeping them in proper uniform and making sure they reported for duty at 6 am each morning.

When I told Sgt. White about my pre-op meeting, he said he already knew about summer maneuvers and was working with the crews on their vehicles.

"Will everything be ready Monday?" I asked. "Hafta wait and see sir. Our vehicles are all soundin' good today in the motor pool, but I wanna hear what they sound like on the road tomorrow," the sergeant said with a chuckle.

The mainstay of the company's supply and mess service was the 'deuce and a half' (ton) truck.

I loved the way Sgt. White covered major issues with his own manner of speaking. The sergeant had solid knowledge and experience behind his easy-going language, and as a black man advancing through the ranks, his astute knowledge and informal disposition got him where he was as a master sergeant.

Walking through the motor pool with Sgt. White, I waved a greeting to my NCOs including my three squad leaders, Sgts. Daniels, Walters, and Jacoby. Like Sgt White – they were career soldiers in for 20 or 30 years and a pension. I admired those guys. They worked in a job they were good at and served their country when we needed them.

The motor pool, with the big tank engines revving up, was like an Indy 500 racing pit. I'd make eye contact and a wave with the guys during my walk-through rather than spoken words, because of the noise. After, I asked Sgt. White to gather the troops for a few minutes before their lunch break.

I began addressing our platoon outside the motor pool, "Men, we're going on summer maneuvers Monday morning , and we need to make sure your vehicles are ready. Sgt. White said you're road-testing them tomorrow. That's good. Let us know if you have any problems. Another important item: 8th Army issued a directive that the C rations and other emergency items stowed in your vehicles must be opened only in the event of a North Korean invasion. That's all. Have a good lunch and keep working on your vehicles this afternoon. Dismissed."

I had my map surveying job to do for Capt. Ramsey, so on the way back from my lunch I stopped at the battalion's S-3 (operations and training) office to pick up the maps of the Chuncheon area. The new S-3, Captain Thompson, was in his office and I asked him about the roads to our op area Monday.

Summer maneuvers featured 7th Cav units in mock battles with the 8th Army infantry, artillery, and aviation units.

"Haven't been there," he said, "but I understand it's mountainous and the roads are through narrow valleys and steep hills. If you're worried about the tracked vehicles getting there, they'll be okay as long as they stay on the roads.

"It was pretty rough going in that eastern sector of Korea during the shooting war," Capt. Thompson continued. "The North Koreans and Chinese had recoilless guns hidden in tunnels high above the roads, and our spotters couldn't see the blasts from their guns to take 'em out. The gooks had a field day with our tanks coming down those narrow roads. It was like shooting ducks in a pond. Pretty grim stuff."

After our chat, I gathered the maps for my survey. "Thanks for the info," I said on leaving his office, and I added. "See you on the road Monday."

I liked Capt. Thompson whom I got to know on a personal basis at the officer's mess. I respected him as an S-3. He was an officer who understood training, quite unlike Capt. Townsend, the former S-3, who was a mother hen about keeping 10-yard intervals and failed to see the need for critiques that were positive and instructive.

Friday was spent in the motor pool and on road tests. "How'd the tests go, sergeant," I asked Sgt. White, about 10:30.

"Vehicles were all in pretty good shape," he said, "but Sgt. Daniels isn't. Seems he had a little too much imbibin' last night, and he's feelin' it this mornin'."

"Sergeant, we've all had hangovers," I said, "Don't let him off the hook. He's a big boy and he'll get through it somehow until lunch." I'd learned not to give in to every complaint from the troops. Sgt Daniels would just have to manage his hangover.

"Sir," said Sgt. White. I noticed a little head shake, like he didn't agree with me. But Sgt. White did as I suggested and told Sgt. Daniel to suck it up. Boozing and hangovers were common, up and down the ranks.

After a full day Friday and half day Saturday for maintenance prep, we were ready and on the road Monday morning, heading east to Chuncheon for the summer maneuvers.

Our vehicles were running pretty well, and I was in the lead jeep with Chou and Sgt. Galant who was with me on my command radio.

Sgt. Galant did a radio check as all 10 platoon vehicles rolled on the dusty road heading east toward the hills and mountains of central Korea. It was a clear summer morning, and the temperature was already up around 85 and getting hotter. We were following the 1st platoon which was churning up a thick dust cloud on the road.

"Watch your interval, Chou. Keep it at 10 yards," I said almost automatically after months of Captain Townsend's scoldings. "Y'sir," he also answered almost automatically. We knew each other well after driving together for eight months.

At our mid-morning break, I checked with Sgt. White. "How the vehicles doing?"

"Lieutenant, they're all' fine. The roads are getting' hillier, but I think all the tracked vehicles will make it," he said.

Sgt. White spoke a little too soon. An hour later, Sgt. Daniel's tank made a sudden lurch to the left at the top of a steep hill, and the tank's tread slipped off its clogged tooth sprocket. It came to an abrupt stop in the middle of the road. Fortunately, Sgt. Daniel's driver was able to move it a little forward and off to the side so the other vehicles in the column could pass by, but Daniel's tank was disabled on the road and close to a steep ditch.

On the way to Chuncheon, Sgt. Daniels' M-48 tank slipped a tank tread at the top of a steep hill and was totally disabled.

"Sgt. Daniels, you got to reengage your tank tread now," I ordered, remembering from my 'Tank Tread Emergency Repair' class I'd taken at Fort Knox.

"I'll try sir, but we're in a bad place on a downward slope. I'll need a level place to reset the track," Daniels replied.

"Well, you don't have a level area. You'll have to do the best you can," I said. "You don't want to be here all day in the hot sun. I'll radio a repair crew to come from HQ but I suggest you first try to fix it yourself. You're in charge."

"Y'sir, I'll do what I can, but it ain't gonna be easy," Daniels complained.

Daniels was 6'3", weighed 220 pounds and could have been a linebacker for the Pittsburgh Steelers. But he was also surprisingly passive for a big man. I was not encouraged by his negative attitude.

He had three others in his crew including a Korean, and I was worried about leaving them. But I also needed to get the rest of my platoon to Chuncheon, so I had to go with the main platoon column and Daniels and crew would have to wait.

Before I left, I made sure we were in radio contact with Daniels. For those unfamiliar with our tanks, our M-48s weighed 45 tons, had a gasoline engine, a 90-millimeter gun, machine guns, a crew of four including a driver, a gunner, a loader, and a tank commander.

An hour later, all the vehicles in Company B arrived at the assigned 8th Army bivouac area. Mess Sgt. Demsey and his truck and crew had already set-up a wonderful Mediterranean-style lunch with cold pasta salad, cold cuts, and iced tea. We were lucky to have Sgt. Demsey with us on that hot Monday day.

All communications on the road were by vehicle radio.

After lunch, the captain had a meeting to set up the afternoon's 8th Army mock battles. I advised him I had to go back to get Sgt. Daniels and his disabled tank. Sgt. White would take over the platoon for the afternoon mock battles with 8th Army.

I radioed to Chief Rembetzi to make sure his HQ maintenance guys were on the way to Daniels. Chou and I then drove back to Daniels' tank. Rembetzi's crew and Chou and I both arrived at Daniels' tank about 3 pm.

What we saw on the disabled tank was Daniels and his crew sitting on top of its turret, eating C-rations out of newly opened cans of baked beans, sausage, spaghetti, meat balls, and fruit cocktail.

They'd obviously found the C-rations stored in their tank that were designated for use *only* in the event of a North Korean invasion.

Sgt. Daniels and his crew climbed down from their perch with their opened C rations to greet us. Daniels' crew then went to work with Sgt. Williams, the chief of the repair crew, in retreading the tank track.

While Rebetzi's and Daniels' crews were working, I pulled Daniels aside and said in a low voice. "Sergeant, what in hell are you thinking opening those stored C-rations?" I started in. "You've disobeyed a standing 8th Army directive."

"Sir, we were hungry, and I didn't know how long we'd be here," Sgt. Daniels said sheepishly under my stern gaze.

"Sergeant! You only missed lunch." I said angrily. "You're supposed to show leadership as a staff sergeant. And you didn't even try to retread the tank's track. You're not looking great in my book today, sergeant!

"And we've got to replace the god damned emergency C rations you opened." I said, "or there'll be hell to pay."

As I confronted Daniels, all I could think about was how to replace the C-ration supplies to keep Daniels from getting in any real trouble.

When we got back to the bivouac at 6, we'd completely missed the mock battle session in the mountains. Sgt. White reported that the infantry again outflanked our tanks on the high ground and the 8th Army controllers had moved our units back all afternoon.

Summer maneuvers was four days of living in the field under combat conditions. The company mess sergeant and his crew providing hot food, three meals a day, for all four days to keep the company fed and in high morale.

Of greater importance, I shared with Sgt. White the problem of Daniels opening his stored C-rations, and the urgent need to replace them before anyone found out.

After a thoughtful pause, Sgt. White replied in a calm voice, "Lieutenant, there are times when its best you leave the details like this to other folks. First, Daniels and his crew should forget their little mistake. Second, nothing more should be spoken about it. I've got it covered from here," Sgt. White said resolutely. And he did.

After we returned to Camp Casey, a completely new C-ration emergency container appeared mysteriously, and was safely stowed inside Sgt. Daniel's M-48. No mention was ever made of Daniel's emergency lunch again.

During live fire, the blast of the 90 mm cannon created a powerful shock wave that hit tank fire observers like a wind tunnel, penetrating their entire body.

10

Company Commander of Company C

Part One

In late July, Col. Edmundson said he wanted to see me in his office after lunch. The last time he asked me to see him was in June when he gave me the wonderful assignment as escort officer of a USO troupe entertaining camps all over Korea.

I knew something was up.

Recently scuttlebutt had it that Captain Tyler, the CO (commanding officer) of Company C, was leaving after his 13-month tour, and HQ had to find another officer from the 7th Cav for the job. All the captains in the unit were already assigned, either as company COs or HQ staff, and there were no replacement captains coming down from 8th Army.

HQ had to pick a 2nd lieutenant from 7th Cav as company commander of Company C. I knew I had the longest date of rank of any of the 2nd lieutenants in the unit, having arrived nine months before, on October 1,1958. But I also knew the Army had strange ways of picking officers for new assignments. I didn't know whether my longest date of rank qualified me for the job. But the colonel asking me to come to his office meant something...

"Sir, reporting as ordered," I saluted.

Returning the salute, the colonel started, "Flagg, I don't know what to do with you. First it was a USO tour in June, and now something new has come up that again involves you," Col. Edmundson said with a serious tone.

"You've not been an altogether bad officer, but you haven't gotten stellar reviews from Ramsey either. Don't you two get along?" the colonel asked.

"Sir. Captain Ramsey and I get along fine, but he wasn't happy with me going on the USO tour during his live fire," I said, not adding that I was still on his shit list.

"Well, enough about Ramsey. You don't seem to have any outstanding demerits in your record," looking at my 201 officer's personnel file open on his desk.

"For the second time, I should have my head examined, Flagg, but you have the longest date of rank of any of our 7th Cav's 2nd lieutenants. I have no choice. I am assigning you as commanding officer of Company C.

After nine months in Korea, Col. Edmundson assigned me as his company commander of Company C, with 120 troops, nine tanks, nine APCs, 16 jeeps, a supply/mess truck, and four acres of company area with 10 Quonsets.

"Lieutenant, this isn't a boondoggle like the USO tour. Running a company is serious business and requires the highest level of commitment and leadership: a helluva lot more than running around with those USO girls and picking lint off your navel. Do you understand?" the colonel asked, looking straight at me.

"Yes sir! "I said affirmatively.

"Fine! Tyler's going back to the States this week, and you're taking over Company C Monday morning. I have some confidence in you, (pause) so don't fuck up and disappoint me," the colonel said without smiling.

"Now go back and kiss and make up with Ramsey. I want you both to start off fresh Monday morning. That's all. Dismissed," the colonel said without smiling.

"Sir," and I saluted. This time I didn't dance out of the HQ as I did in June with the USO tour. Running Company C was going to be a challenge, one I hoped I was up for.

Flagg was the commanding officer of Company C, but the company's smooth operation was the work of the experienced company career sergeants.

That weekend I was especially anxious about my new assignment. At our usual poker game on Saturday afternoon, I asked the colonel a question before the other players sat down, "Sir, should I go to the company over the weekend and check it out?

He thought for a moment and shook his head, no. "You got to let it sink in a little. They're getting a new boss on Monday. Let 'em relax and enjoy the weekend."

Colonel Edmundson's excellent advice that Saturday gave me a good start with my new officers and NCOs that Monday. They had the whole weekend to get all the scuttlebutt about me; I played poker with the colonel; I wasn't a huge pain in the ass and probably would be okay. though not up to Capt. Tyler's relaxed high standards.

Monday morning, Capt. Tyler's driver Sam was at the officer's mess at 7:30 to pick me up and the two other lieutenants assigned to Company C and now working for me. Tom Clarkson was my friend from B Company (who came a week after me and played on the Camp Casey baseball team). And Danny St. Clair, a gentle giant who had a good relationship with everyone and could speak better Korean than almost everyone except the Koreans.

The top sergeant of Company C was Master Sgt. Jerold Leonard, who met me at the company HQ. Leonard knew his job cold: running the 120-man company; keeping the other NCOs in line; and most importantly for me, making the commanding officer look good. No wonder Capt. Tyler always seemed relaxed. With Leonard on the job, he could be relaxed, unlike Capt. Ramsey of B Company, who always seemed anxious.

Everyone called Sgt. Leonard 'Top,' and it was a deserved title (for top sergeant). He was prepared for me, as Col. Edmundson had predicted, and made my transition orderly and easy.

There was one surprise in assuming my command for which I wasn't prepared. When an officer is appointed a company commander, he must sign for every item of company property. The battalion S-1, Major Dempsey, was waiting for me when I arrived at the company HQ. He was Col. Edmundson's administrative officer and had come with his briefcase and a complete list of U.S. Government property for the company, a 22-page inventory together with a receipt form for me to sign.

The list included everything within the four acres of Company C - six tanks, six APCs, 16 jeeps, a supply/mess truck, all the motor pool equipment, all the arms room weapons including 150 rifles, 10 pistols, 15 machine guns, and ammunition for each, all the kitchen equipment, and all the supply room stores. The list also included the 10 Quonsets, barracks bedding, bathroom toilets and showers.

The major suggested I look over the inventory and sign my name and rank on the separate receipt form, while he waited. "Don't worry," he said in a matter-of-fact manner. "It's just a routine, lieutenant. Every officer taking over a company must sign for the company's property."

"How much is this stuff worth?" I asked incredulously, thumbing through the inventory.

"Ya' know, I never bothered to add it up, but it's probably 'round $15 or $20 million," he said. "But don't worry. It's only Monopoly money. No one's really expects you to pay for it," he said. "Unless you lose it!"

I glanced at him and saw a little smirk, like he had just said something funny. To him, it was just a joke, and I should sign it and forget it. Which I did, realizing the signing ritual was just another Army game. I wasn't going to Leavenworth Army prison for losing a tank. I was still learning about the Army.

Part Two

My Company C command only lasted a short three months until my 13-month tour was up the end of October. During that time, two incidents exemplify my experience commanding an armored unit with 120 troops – half of them from the Korean Army.

The first incident occurred shortly after I took over.

Company C had a live fire exercise scheduled for July, and it had been scheduled well before I took over the company. The live fire date was the second week I was on the job. It was the same live fire

I'd missed a month earlier in Company B, when the colonel gave me my USO job.

Our 90 mm tank guns required firing at least once a year and live fire served that purpose. The only gunnery range for tanks was a valley 30 miles west of Camp Casey. We needed four days for the live fire trip: a day to get there and set up sleeping tents and the company kitchen; two days of live fire with each tank; and a day to return.

While I was CO and in charge, the real leaders were the experienced NCOs led by Top himself, Master Sgt. Leonard. The NCOs were the crew chiefs in each tank, APC, and jeep in our caravan. They were also the camp honchos in setting up the large squad sleeping tents for the troops, the kitchen service and chow line, and everything else, from the latrine detail to camp security.

What surprised me most was a new perk I had as the CO of Company C. Top had set up my own personal sleeping tent, with cot, mattress, pillows, sheets, pillowcases, and a lantern. When I asked him if Capt. Tyler received the same VIP treatment, he just looked at me, and almost disdainfully said, "We *always* take care of our COs."

The next day each tank rolled into its firing position on the range with targets set at 2000 yards, over a mile away. A mountain range was a quarter mile behind the targets to safely catch the overs of each HE (high explosive shell), AP (armor piercing shell), and other shells traveling at a muzzle velocity of 3,200 feet per second (3 times the speed of sound).

The firing range observers, including me, used binoculars to score the tank's accuracy while standing on the tank, next to the turret for a better view of the target over a mile away. With each cannon blast, the noise level was beyond hearing. What we felt instead was an incredible shock wave that penetrated through our entire bodies like a gigantic wind tunnel blast.

Most of the NCOs on the range, and all of the officers, chose not to put cotton in their ears in the belief that it was unsoldierly and would set a bad example for the troops. How wrong we were. Recent Army studies have confirmed that hearing loss and even brain damage is prevalent among gun crews today. Those crews are almost identical to the crews of our live fire exercise in Korea. I have hearing loss today that I believe started at live fire 67 years ago.

Company C's live fire was a success. Col. Edmundson drove over the first day of firing and was pleased with what he saw. "Lieutenant, your camp is well set up and your firing operation is excellent. You're lucky to have inherited Capt. Tyler's well-run company. I hope you keep it that way," was his only comment. No mention was made of my leadership. The colonel knew the company was driven by experienced NCOs who collectively made the company (and me) look good.

I *was* lucky to have inherited Tyler's company. I thought live fire would be difficult, but it proved to be a smooth operation. The NCOs and the officers pitched in to make live fire and my first command, successful.

The second incident involved the colonel.

Company C was allotted a hundred bags of cement in August to harden walkways that were dirt paths in front of the enlisted men's quarters. When it rained, those paths became enormous puddles, yet they were in constant use by troops coming and going from their barracks to the mess hall and the motor pool.

Top and the NCOs were enthusiastic about the walkway project, and I gave them free rein to use their borrowing and scrounging skills to get the project underway.

First, they borrowed a portable cement mixer from HQ. Next, they found a source for sand and gravel a mile outside of Camp Casey and used our deuce-and-a-half to haul the stuff back to our mixing site. Then, they scrounged lumber from an unnamed source (maybe they stole it) and constructed a movable 6' x 4' sidewalk form to pour the mixed concrete until it set.

The project progressed perfectly, setting a new sidewalk in 6' x 4' increments every two or three days after the cement had been set. But August was an extraordinarily hot month, and the Korean sun blistered 90-degree heat almost every day. There was a 7th Cav directive stating that troops must wear fatigue pants and shirts buttoned to the wrists to prevent mosquito bites and other diseases.

One Thursday afternoon the men were busy with the sidewalk project, mixing the next batch of concrete.

Company C was allotted a hundred bags of cement for new barracks cement walkways. The colonel intervened and almost threatened my command .

The temperature was over 90 degrees with no wind, and I was supervising. Sgt. Musgrave, the crew chief that day, came over and asked, "Lieutenant, can't we take off our shirts off today. It's stifling hot, and the job will go a lot faster if we're not boiling in these clothes."

"Sure," I said, "Take 'em off," not thinking about anything more than helping the crew.

Shortly after, Col Edmundson rolled up in his jeep to see how our sidewalk project was going. When he walked up the hill to the sidewalk, he suddenly took off his sunglasses and scowled. His face turned red. "Lieutenant! I need to see you in private. Now! Down by my jeep!"

The colonel could barely get the words out of his mouth. Down by his jeep, he excused his driver and lit into me.

"Flagg, don't you understand what standing orders mean?" he almost shouted, angrily shaking his head. "Shirts stay on in the

7th Cav! Preventing mosquito bites and malaria should be a CO's number one priority!"

'Yes Sir." I offered.

"Don't 'Yes Sir' me. You disappoint me! I have no choice but to put a 'failure to follow standing orders' in your permanent 201. Now get back up there and get their shirts on! That's all, dismissed."

The colonel then slowly put his sunglasses back on, still shaking his head, waved to his driver that he was ready to go, and they drove off.

I had never seen him so angry. I knew he would probably relieve me of my command, and I was also embarrassed to be chewed out in front of my troops. Most of all, I was angry at myself for trying to do the right thing but screwing up.

Meanwhile the sidewalk cement crew on the hill had stopped work and stood gawking at a lieutenant being chewed out at by the colonel.

When I walked back up the hill, they didn't say a word, but it was clear when they looked me that they wanted to know what had happened.

"At ease, men. What you just saw (pause) was how to really piss off a lieutenant colonel. You gotta get your shirts back on. It's a standing 7th Cav order to keep shirts on to prevent mosquito bites and disease. My mistake in letting you take 'em off."

"Sir," Sgt. Musgrave said walking over to me after about 5 minutes. He spoke in a conspiratorial whisper, "We know you were trying to do the right thing." He then went back to pouring concrete in the 6' x 4' forms with his crew.

The colonel never mentioned the incident again, never relieved me of command, and never put anything in my 201 file.

I was still learning about the Army.

Epilogue

When I finished my tour of duty at Camp Casey at the end of October 1959, I left Korea on a high note. I had become the commanding officer of a combat company,

This book follows the chronology of my 13-months in Korea as a new 2nd lieutenant. At the beginning of my tour, I could barely manage a 7-man work detail. At the end, I was a company commander.

I was competent enough after 8 months to be given the assignment of escorting a USO troupe throughout South Korean to entertain troops at 11 different Army bases. I was their manager for the tour and traveled with them for two weeks in a two-and-a-half ton army truck over the entire 240-mile route north, from Busan on the Southeast coast of Korea, to Camp Casey, 10 miles from the DMZ.

My last assignment at the 7 th Cav was the culmination of all my training in Korea. I was assigned to Company C as their commanding officer, a combat reconnaissance unit with 120 troops. The USO tour and my commanding officer assignment were two highlights of my tour that set me apart from other 2nd lieutenants at the 7 th Cav.

I was fortunate during my tour to have two outstanding platoon sergeants assigned to my platoon. MSgt. Temple and MSgt. White were mentors who taught me by example how to lead men. My two battalion commanders of the 7 th Cav were also important mentors in providing special assignments in developing my leadership skills.

Lt.Col. Stephenson, during my first months on tour, found assignments that built a sense of confidence and purpose as a new officer. Lt.Col. Edmundson, later, found wonderful special assignments for me as an experienced officer that including my USO tour and my commanding officer assignment at Company C.

All those mentors contributed to my development as an officer and resulted in my time in Korea being a memorable experience.

Read Author Russell Flagg's Other Books

Go Behind the Scenes in the Blockbuster World of Trade Shows

. Discover secrets you may not know about the multi-billion-dollar Trade Show world...

. the British Royal who created the first Great Exhibition in 1851 in London

. the show that's now the largest annual US trade show

. where Home Depot, Lowes and Ace Hardware go to stock shelves with hardware merchandise

. the trade show wizard who brought shows to over 10 American markets

. the financial show that began in NY and quickly spread to other countries

Kidnapping In Pittsbuwgh

Flagg's memoir takes readers to wartime Pittsburgh as seen through the eyes of a seven-year-old. Pittsburgh in the 1940s had streetcars, booming factories, and working families managing, with a world war raging in Europe and Asia.

"Kidnapping," was the families nickname for the art of convincing local mothers to purchase a photo portrait of their children. Selling those photos was the family's source of income during the years the author's parents worked for Alan Studio in Pittsburgh.

About the Author

Russ celebrates his 89th birthday with too many candles.

Russ Flagg lives in Narragansett, Rhode Island with his artist and graphic designer wife, Holley Flagg. He attended Syracuse University and after four years of ROTC, he graduated with a 2nd lieutenant's commission in the US Army.

After his basic officer training at Fort Knox, Flagg's first assignment was a 13-month duty tour in South Korea with the 7th Cavalry reconnaissance battalion. After eight months there, he was assigned as the escort officer for a USO troupe touring 11 Army camps throughout South Korea. At the end of his duty tour, he was assigned as company commander of a combat unit with 120 troops.

This Army experience was unique for a young officer, and the 13-month tour provided the leadership skills he needed to take on those important assignments.

Flagg states that "escorting the USO tour all over Korea, and commanding a company were highlights of my 13-month tour. I was also lucky not to be in a shooting war in Korea, or in the Vietnam war later. My tour of duty in Korea was memorable and why I wrote this journal."

Made in United States
North Haven, CT
04 May 2026

10863440R00063